Transparent User Authentication

Nathan Clarke

Transparent User Authentication

Biometrics, RFID and Behavioural Profiling

Springer

Nathan Clarke
Centre for Security, Communications &
Network Research (CSCAN)
Plymouth University
Drake Circus
PL4 8AA Plymouth
United Kingdom
N.Clarke@plymouth.ac.uk

ISBN 978-0-85729-804-1 e-ISBN 978-0-85729-805-8
DOI 10.1007/978-0-85729-805-8
Springer London Dordrecht Heidelberg New York

British Library Cataloguing in Publication Data
A catalogue record for this book is available from the British Library

Library of Congress Control Number: 2011935034

Printed on acid-free paper

Springer is part of Springer Science+Business Media (www.springer.com)

Preface

The world of user authentication is focussed upon developing technologies to solve the problem of point-of-entry identity verification required by many information systems. Unfortunately, authentication approaches; secret knowledge, token and biometric, all fail to provide universally strong user authentication – with various well-documented failings existing. Moreover, existing approaches fail to identify the real information security risk. Authenticating users at point-of-entry, and failing to require re-authentication of the user during the session provides a vast opportunity for attackers to compromise a system. However, forcing users to continuously re-authenticate to systems is cumbersome and fails to take into account the human factors of good security design, in order to ensure good levels of acceptability. Unfortunately, within this context, the need to authenticate is increasing rather than decreasing, with users interacting and engaging with a prolific variety of technologies from PCs to PDAs, social networking to share dealing, and Instant Messenger to Texting. A re-evaluation is therefore necessary to ensure user authentication is relevant, usable, secure and ubiquitous.

The book presents the problem of user authentication from a completely different standpoint to current literature. Rather than describing the requirements, technologies and implementation issues of designing point-of-entry authentication, the text introduces and investigates the technological requirements of implementing transparent user authentication – where authentication credentials are captured during a user's normal interaction with a system. Achieving transparent authentication of a user ensures the user is no longer required to provide explicit credentials to a system. Moreover, once authentication can be achieved transparently, it is far simpler to perform continuous authentication of the user minimising user inconvenience and improving the overall level of security. This would transform current user authentication from a binary point-of-entry decision to a continuous identity confidence measure. By understanding the current confidence in the identity of the user, the system is able to ensure that appropriate access control decisions are made – providing immediate access to resources with high confidences and requiring further validation of a user's identity with low confidences.

Part I begins by reviewing the current need for user authentication – identifying the current thinking on point-of-entry authentication and why it falls short of providing real and effective levels of information security. Chapter 1 focuses upon the current implementation of user authentication and places the role of authentication within the wider context of information security. The fundamental approaches to user authentication and their evolutions are introduced. Chapter 2 takes an opportunity to review the need for user authentication through an examination of the history of modern computing. Whilst authentication is key to maintaining systems, it is frequently overlooked and approaches are adopted that are simply not fit for purpose. In particular, the human aspects of information security are introduced looking at the role the user plays in providing effective security. The final chapter in Part I investigates the role of user authentication in modern systems, what it is trying to achieve and more importantly, if designed correctly, what it could achieve. A discussion on the applicability of utilising risk assessment and how continuous authentication would function are described.

Part II is focussed upon the authentication approaches and providing an in-depth analysis of how each operates. Chapter 4 takes each of the three fundamental approaches in turn and discusses the various implementations and techniques available. The chapter presents how each of the systems works and identifies key attacks against them. Having thoroughly examined traditional authentication, Chap. 5 investigates transparent authentication approaches. Supported by current literature and research, the chapter details how transparent authentication can be accomplished and the various technological barriers that currently exist. Taking the concept of transparent and continuous authentication further, Chap. 6 discusses multimodal authentication. The chapter details what multimodal authentication is, what methods of fusion exist and its applicability in this context. The final chapter in Part II, describes the standardisation efforts currently underway in the field of biometrics. Only through standardisation will widespread vendor-independent multimodal systems be able to exist.

Part III examines the wider system-specific issues with designing large-scale multimodal authentication systems. Chapters 8 and 9 look at the theoretical and practical requirements of a system and discuss the limitations and advantages such a system would pose. Obviously, with increasing user authentication and use of biometrics, the issue of privacy arises and Chap. 9 focuses upon the need to ensure privacy and the human factors of acceptability and perception. The book concludes with a look into the future of user authentication, what the technological landscape might look like and the effects upon the people using these systems.

Acknowledgements

For the author, the research presented in this book started at the turn of the century and represents a decade of research undertaken. During this time, a number of M.Sc. and Ph.D. students have contributed towards furthering aspects of the research problem, and thanks are due in no small part to all of them. Many of the examples used in this book and their experimental findings are due to them. Specific thanks are due to Fudong Li for his work on behavioural biometrics, Christopher Hocking for conceptualising the Authentication Aura and Sevasti Karatzouni for her work on the implementation and evaluation of early prototypes, and in particular her invaluable contribution in Chap. 9.

The initial concept of performing authentication transparently needs to be credited to my colleague and mentor Prof. Steven Furnell. It was due to his creativity that the concept was initially created. He also needs to be credited with the guiding hand behind much of the work presented in this book. It is only through his encouragement and support that this book was made possible.

The reviewing and critiquing of book chapters is a time-consuming and arduous task and thanks are due to Christopher Bolan in particular who gave a considerable amount of personal time examining the manuscript. Along with others, I appreciate all the time, patience and advice they have given.

Thanks are also due to all the companies and organisations that have funded aspects of this research over the past 10 years. Specifically, thanks are due to the Engineering and Physical Sciences Research Council (EPSRC), Orange Personal Communications Ltd, France-Telecom, the EduServ Foundation and the University of Plymouth.

I would also like to thank Simon Rees from Springer for his initial and continued support for the book, even when timelines slipped and additions were made that led to the text being delayed. I would also like to thank all the staff at Springer that have helped in editing, proofing and publishing the text.

Final thanks are due to my wife, Amy, who has had to put up with countless evenings and weekends alone whilst I prepared the manuscript. She was the inspiration to write the book in the first place and I am appreciative of all the support, motivation and enthusiasm she provided. Thankfully, this did not put her off marrying me!

Contents

Part II Authentication Approaches

List of Figures

List of Tables

Part I
Enabling Security Through User Authentication

Chapter 1
Current Use of User Authentication

1.1 Introduction

Information security has become increasingly important as technology integrates into our everyday lives. In the past 10 years, computing-based technology has permeated every aspect of our lives from desktop computers, laptops and mobile phones to satellite navigation, MP3 players and game consoles. Whilst the motivation for keeping systems secure has changed from the early days of mainframe systems and the need to ensure reliable audits for accounting purposes, the underlying requirement for a high level of security has always been present.

Computing is now ubiquitous in everything people do – directly or indirectly. Even individuals who do not engage with personal computers (PCs) or mobile phones still rely upon computing systems to provide their banking services, to ensure sufficient stock levels in supermarkets, to purchase goods in stores and to provide basic services such as water and electricity. In modern society there is a significant reliance upon computing systems – without which civilisation, as we know it, would arguably cease to exist.

As this reliance upon computers has grown, so have the threats against them. Whilst initial endeavours of computer misuse, in the late 1970s and 1980s, were largely focused upon demonstrating technical prowess, the twenty-first century has seen a significant focus upon attacks that are financially motivated – from botnets that attack individuals to industrial espionage. With this increasing focus towards attacking systems, the domain of information systems security has also experienced increasing attention.

Historically, whilst increasing attention has been paid to securing systems, such a focus has not been universal. Developers have traditionally viewed information security as an additional burden that takes significant time and resources, detracting from developing additional functionality, and with little to no financial return. For organisations, information security is seen as rarely driving the bottom line, and as such they are unmotivated to adopt good security practice. What results is a variety of applications, systems and organisations with a diverse set of security polices and

N. Clarke, *Transparent User Authentication: Biometrics, RFID and Behavioural Profiling*, DOI 10.1007/978-0-85729-805-8_1, © Springer-Verlag London Limited 2011

levels of adoption – some very secure, a great many more less so. More recently, this situation has improved as the consequences of being successfully attacked are becoming increasingly severe and public. Within organisations, the desire to keep intellectual property, regulation and legislation is a driving factor in improving information security. Within the services and applications, people are beginning to make purchasing decisions based upon whether a system or application is secure; driving developers to ensure security is a design factor.

Authentication is key to providing effective information security. But in order to understand the need for authentication it is important to establish the wider context in which it resides. Through an appreciation of the domain, the issues that exist and the technology available, it is clear why authentication approaches play such a pivotal role in securing systems. It is also useful to understand the basic operation of authentication technologies, their strengths and weaknesses and the current state of implementation.

1.2 Basics of Computer Security

The field of computer security has grown and evolved in line with the changing threat to landscapes and the changing nature of technology. Whilst new research is continually developing novel mechanisms to protect systems, the fundamental principles that underpin the domain remain unchanged. Literature might differ a little on the hierarchy of all the components that make up information security; however, there is an agreement upon what the key objectives or goals are. The three aims of information security are Confidentiality, Integrity and Availability and are commonly referred to as the CIA triad. In terms of information, they can be defined as follows:

- Confidentiality refers to the prevention of unauthorised information disclosure. Only those with permission to read a resource are able to do so. It is the element most commonly associated with security in terms of ensuring the information remains secret.
- Integrity refers to ensuring that data are not modified by unauthorised users/processes. Integrity of the information is therefore maintained as it can be changed only by authorised users/processes of a system.
- Availability refers to ensuring that information is available to authorised users when they request it. This property is possibly the least intuitive of the three aims but is fundamental. A good example that demonstrates the importance of availability is a denial of service (DoS) attack. This attack consumes bandwidth, processing power and/or memory to prevent legitimate users from being able to access a system.

It is from these three core goals that all information security is derived. Whilst perhaps difficult to comprehend in the first instance, some further analysis of the root cause of individual attacks does demonstrate that one or more of the three

Table 1.1 Computer attacks affecting CIA

Attack	Security goal		
	Confidentiality	Integrity	Availability
(Distributed) Denial of service			✓
Hacker	✓	✓	
Malicious software (e.g. Worms, Viruses, Trojans)	✓	✓	✓
Phishing	✓		
Rootkit	✓	✓	
Social engineering	✓		
Spam			✓

security goals are being affected. Consider, for instance, the role of the computer virus. Fundamentally designed to self-replicate on a computer system, the virus will have an immediate effect upon the availability of system resources, consuming all the memory and processing capacity. However, depending upon the severity of the self-replicating process, they can also have an effect upon the integrity of the data stored. Viruses also have some form of payload, a purpose or reason for existing, as few are non-malignant. This payload can vary considerably in purpose, but more recently Trojans have become increasingly common. Trojans will search and capture sensitive information and relay it back to the attacker, thereby affecting the confidentiality of the information. To illustrate this further, Table 1.1 presents a number of general attacks and their effect upon the security objectives.

In addition to the goals of information security, three core services support them. Collectively referred to as AAA, these services are Authentication, Authorisation and Accountability. In order to maintain confidentiality and integrity, it is imperative for a system to establish the identity of the user so that the appropriate permissions for access can be granted, without which anybody would be in a position to read and modify information on the system. Authentication enables an individual to be uniquely identified (albeit how uniquely is often in question!) and authorisation provides the access control mechanism to ensure that users are granted their particular set of permissions. Whilst both authentication and authorisation are used for proactive defence of the system (i.e. if you don't have a legitimate set of authentication credentials you will not get access to the system), accountability is a reactive service that enables a system administrator to track and monitor system interactions. In cooperation with authentication, a system is able to log all system actions with a corresponding identity. Should something have gone amiss, these logs will identify the source and effect of these actions. The fact that this can only be done after an incident makes it a reactive process. Together, the three services help maintain the confidentiality, integrity and availability of information and systems.

Looking at security in terms of CIA and AAA, whilst accurate, paints a very narrow picture of the information security domain. Information security is not merely about systems and technical controls utilised in their protection. For instance, whilst authentication does indeed ensure that access is only granted to a legitimate identity, it does not consider that the authentication credential itself might be

Fig. 1.1 Facets of information security

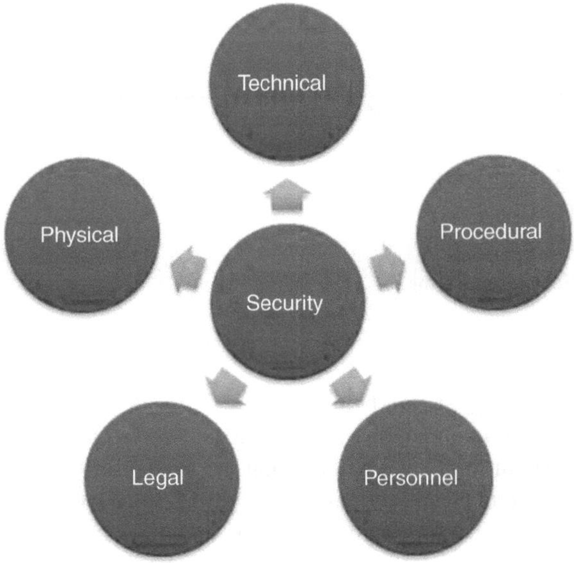

compromised through human neglect. Therefore, any subsequent action using that compromised credential will have an impact upon the confidentiality and integrity of the information. Furnell (2005) presents an interesting perspective on information security, in the form of a jigsaw puzzle comprising the five facets of information: technical, procedural, personnel, legal and physical (as illustrated in Fig. 1.1). Only when the jigsaw is complete and all are considered together can an organisation begin to establish a good information security environment. A lack of considering any one element would have serious consequences on the ability to remain secure.

Whilst the role of the technical facet is often well documented, the roles of the remaining facets are less so. The procedural element refers to the need for relevant security processes to be undertaken. Key to these is the development of a security policy, contingency planning and risk assessment. Without an understanding of what is trying to be achieved, in terms of security, and an appreciation that not all information has the same value, it is difficult to establish what security measures need to be adopted. The personnel element refers to the human aspects of a system. A popular security phrase, 'security is only as strong as its weakness link', demonstrates that a break in only one element of the chain would result in compromise. Unfortunately, literature has demonstrated that the weakest link is frequently the user. The personnel element is therefore imperative to ensure security. It includes all aspects that are people-related, including education and awareness training, ensuring that appropriate measures are taken at recruitment and termination of employment and maintaining a secure behaviour within the organisation. The legal element refers to the need to ensure compliance with relevant legislation. An increased focus upon legislation from many countries has resulted in significant controls on how organisations use and store information. It is also important for an organisation to comply with legislation in all countries in which it operates. The volume of legislation is also growing, in part to better protect systems. For example, the following are a sample of the laws that would need to be considered within the UK:

- Computer Misuse Act 1990 (Crown 1990)
- Police and Justice Act 2006 (included amendments to the Computer Misuse Act 1990) (Crown 2006)
- Regulation of Investigatory Powers Act 2000 (Crown 2000a)
- Data Protection Act 1998 (Crown 1998)
- Electronic Communication Act 2000 (Crown 2000b)

In addition to legislation, the legal element also includes regulation. Regulations provide specific details on how the legislation is to be enforced. Many regulations, some industry-specific and others with a wider remit, exist that organisations must legally ensure they comply against. Examples include:

- The US Health Insurance Portability and Accountability Act (HIPAA) requires all organisations involved in the provision of US medical services to conform to its rules over the handling of medical information.
- The US Sarbanes-Oxley Act requires all organisations doing business in the US (whether they are a US company or not) to abide by the act. Given many non-US companies have business interests in the US, they must ensure they conform to the regulation.

Finally, the physical element refers to the physical controls that are put into place to protect systems. Buildings, locked doors and security guards at ingress/egress points are all examples of controls. In the discussion thus far, it has almost been assumed that these facets related to deliberate misuse of systems. However, it is in the remit of information security to also consider accidental threats. With respect to the physical aspect, accidental threats would include the possibility of fire, floods, power outages or natural disasters. Whilst this is conceivably not an issue for many companies, for large-scale organisations that operate globally, such considerations are key to maintaining availability of systems. Consider, for example, what would happen to financial institutions if they did not consider these aspects to be appropriate. Not only would banking transaction data be lost, access to money would be denied and societies would grind to a stop. The banking crisis of 2009/2010 where large volumes of money were lost on the markets, which consequently caused a global recession, is a good example of the essential role these organisations play in daily life and the impact they have upon individuals.

When considering how best to implement information security within an organisation, it is imperative to ensure an organisation knows what it is protecting and why. Securing assets merely for the sake of securing them is simply not cost-effective and paying £1,000 to protect an asset worth only £10 does not make sense. To achieve this, organisations can undertake an information security risk assessment. The concept of risk assessment is an understanding of the value of the asset needing protection, the threats against the asset and the likelihood or probability that the threat would become a reality. As illustrated in Fig. 1.2, the compromise of the asset will also have an impact upon the organisation and a subsequent consequence. Once a risk can be

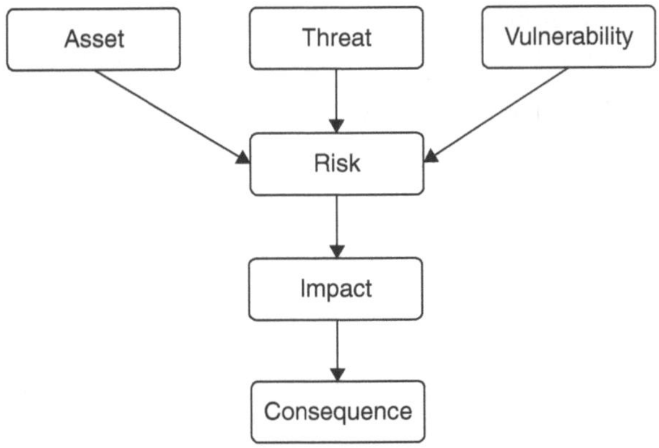

Fig. 1.2 Information security risk assessment

quantified, it is possible to consider the controls and countermeasures that can be put into place to mitigate the risk to an acceptable level.

For organisations, particularly smaller entities, a risk assessment approach can be prohibitively expensive. Baseline standards, such as the ISO27002 Information Security Code of Practice (ISO 2005a), provide a comprehensive framework for organisations to implement. Whilst this does not replace the need for a solid risk assessment, it is a useful mechanism for organisations to begin the process of being secure without the financial commitment of a risk assessment. The process of assessing an organisation's security posture is not a one-off process, but as Fig. 1.3 illustrates is a constantly reoccurring process, as changes in policy, infrastructure and threats all impact upon the level of protection being provided.

The controls and countermeasures that can be utilised vary from policy-related statements of what will be secured and who is held responsible to technical controls placed on individual assets, such as asset tagging to prevent theft. From an individual system perspective, the controls you would expect to be included are an antivirus, a firewall, a password, access control, backup, intrusion detection or prevention system, anti-phishing and anti-spam filters, spyware detection, application and operating system (OS) update utility, a logging facility and data encryption. The common relationship between each countermeasure is that each and every control has an effect upon one or more of three aims of information security: confidentiality, integrity or availability.

The effect of the controls is to eliminate or, more precisely, mitigate particular attacks (or sets of attacks). The antivirus provides a mechanism for monitoring all data on the system for malicious software, and the firewall blocks all ports (except for those required by the system), minimising the opportunity for hackers to enter into the system. For those ports still open, an Intrusion Detection System is present, monitoring for any manipulation of the underlying network protocols. Finally at the application layer, there are application-specific countermeasures, such as anti-spam and anti-phishing, that assist in preventing compromise of those services. As illustrated in Fig. 1.4, these countermeasures are effectively layered, providing a 'defence in depth' strategy, where any single attack needs to compromise more than one security control in order to succeed.

Fig. 1.3 Managing information security

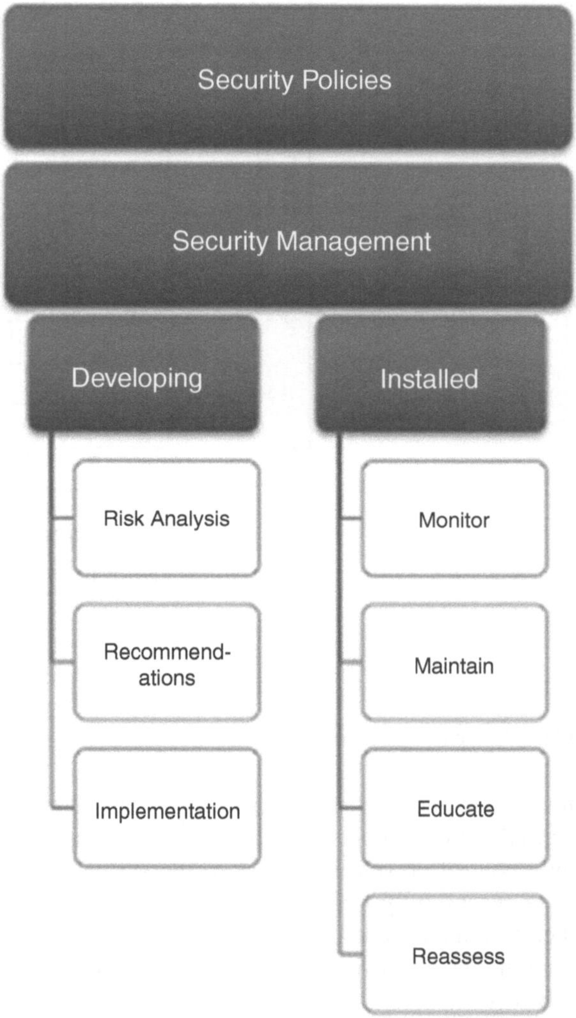

An analysis of Fig. 1.4 also reveals an overwhelming reliance upon a single control. From a remote, Internet-based attack perspective, the hacker has a number of controls to bypass, such as the firewall and intrusion detection system. A target for the attacker would therefore be to disable the security controls. In order to function, these controls are configurable so that individual users can set them up to meet their specific requirements. These configuration settings are secured from misuse by an authentication mechanism. If the firewall software has any software vulnerability, a hacker can take advantage of the weakness to obtain access to the firewall. Once the compromise is successful, the hacker is able to modify the firewall access control policy to allow for further attacks. Similar methods can be applied to the other countermeasures. For instance, switching off or modifying the antivirus is a common strategy deployed by malware. If the system is set up

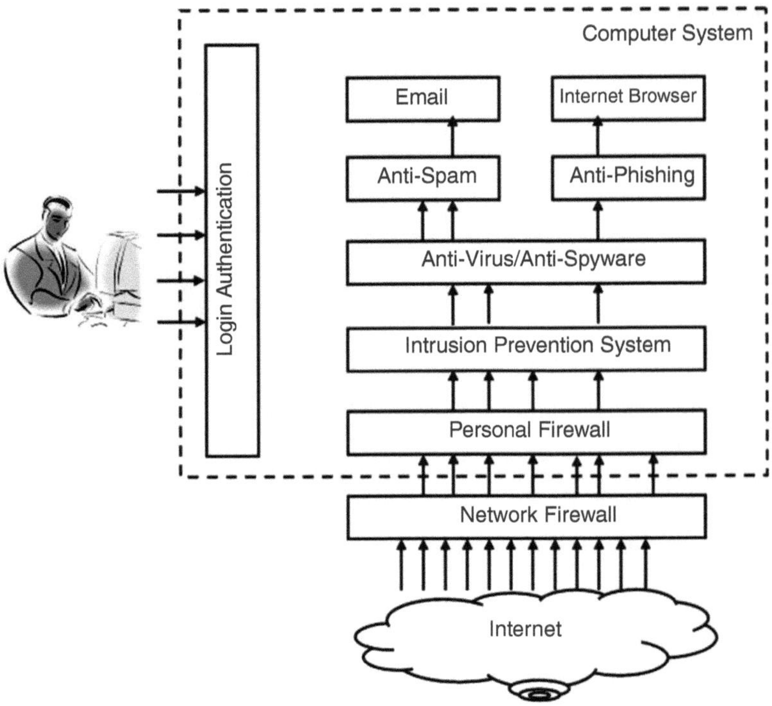

Fig. 1.4 Typical system security controls

for remote access, the hacker needs to only compromise the authentication credentials to obtain access to the system. From a physical attack perspective, the only control preventing access to the system is authentication – assuming they have successfully bypassed the physical protection (if present). Authentication therefore appears across the spectrum of technical controls. It is the vanguard in ensuring the effective and secure operation of the system, applications and security controls.

1.3 Fundamental Approaches to Authentication

Authentication is key to maintaining the security chain. In order for authorisation and accountability to function, which in turn maintain confidentiality, integrity and availability, correct authentication of the user must be achieved. Whilst many forms of authentication exist such as passwords, personal identification numbers (PINs), fingerprint recognition, one-time passwords, graphical passwords, smartcards and Subscriber Identity Modules (SIMs), they all fundamentally reside within one of three categories (Wood 1977):

- Something you know
- Something you have
- Something you are

Something you know refers to a secret knowledge–based approach, where the user has to remember a particular pattern, typically made up of character and numbers. Something you have refers to a physical item the legitimate user has to unlock the system and is typically referred to as a token. In non-technological applications, tokens include physical keys used to unlock the house or car doors. In a technological application, such as remote central locking, the token is an electronic store for a password. Finally, something you are refers to a unique attribute of the user. This unique attribute is transformed into a unique electronic pattern. Techniques based upon something you are, are commonly referred to as biometrics.

The password and PIN are both common examples of the secret-knowledge approach. Many systems are multi-user environments and therefore the password is accompanied with a username or claimed identity. Whilst the claimed identity holds no real secrecy, in that a username is relatively simple to establish, both are used in conjunction to verify a user's credentials. For single-user systems, such as mobile phones and personal desktop assistance (PDA), only the password or PIN is required.

The strength of the approach resides in the inability for an attacker to successfully select the correct password. It is imperative therefore that the legitimate user selects a password that is not easily guessable by an attacker. Unfortunately, selecting an appropriate password is where the difficulty lies. Several attacks from social engineering to brute-forcing can be used to recover passwords and therefore subsequently circumvent the control. Particular password characteristics make this process even simpler to achieve. For instance, a brute-force attack simply tries every permutation of a password until the correct sequence is found. Short passwords are therefore easier to crack than long passwords. Indeed, the strength of the password is very much dependent upon ensuring that the number of possible passwords or the password space is so large that it would be computationally difficult to brute-force a password in a timely fashion. What defines timely is open to question depending upon the application. If it is a password to a computer system, it would be dependent on how frequently the password is changed – for instance, a password policy stating that passwords should change monthly would provide a month to a would-be attacker. After that time, the attacker would have to start again. Longer passwords therefore take an exponentially longer time to crack. Guidelines on password length do vary with password policies in the late 1990s, suggesting that eight characters was the minimum. Current attacks such as Ophtcrack (described in more detail in Sect. 4.2.3) are able to crack 14-character random passwords in minutes (Ophcrack 2011).

Brute-forcing a password (if available) represents the most challenging attack for hackers – and that is not particularly challenging if the password length is not sufficient. However, there are even simpler attacks against trying every permutation. This attack exploits the user's inability to select a completely random password. Instead they rely upon words or character sequences that have some meaning. After all, they do have to remember the sequence and truly random passwords are simply not easy to remember. A typical example of this is a word with some meaning, take 'luke' (my middle name) appended to a number '23' (my age at the time) – *luke23*. Many people perceive this to be a strong password, as it does not rely upon a single

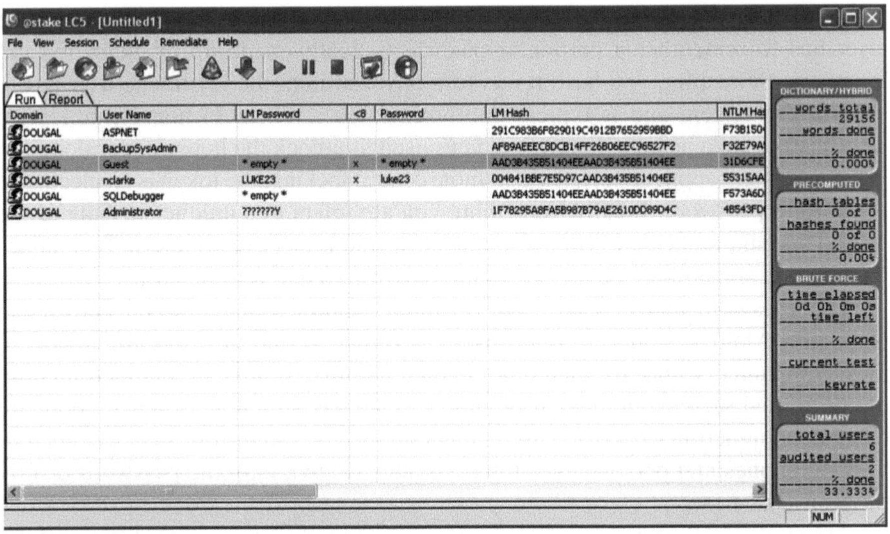

Fig. 1.5 Lophcrack software

dictionary word. However, software such as AccessData's Password Recovery Toolkit (AccessData 2011) and Lophcrack (Security Focus 2010) have a process for checking these types of sequence prior to the full brute force. Figure 1.5 illustrates Lophcrack breaking this password, which was achieved in under 2 min.

It is worth noting that these types of attack are not always possible and do assume that certain information is available and accessible to an attacker. In many situations where passwords are applied this is not the case. In those situations, as long as the three-attempt rule is in place (i.e. the user gets three attempts to login, after which the account is locked) these types of brute-forcing attacks are not possible. However, because of other weaknesses in using passwords, an attacker gaining access to one system can also frequently provide access to other systems (where brute-forcing was not possible) as passwords are commonly shared between systems. If you consider the number of systems that you need to access, it soon becomes apparent that this is not an insignificant number and is typically increasing over time. For instance, a user might be expected to password protect:

- Work/home computer
- Work network access
- Work email/home email
- Bank accounts (of which he/she could have many with different providers – mortgage, current, savings, joint account)
- Paypal account
- Amazon account
- Home utilities (gas, electricity, water services all maintain online accounts for payment and monitoring of the account)
- Countless other online services that require one to register

It is simply not possible for the average user to remember unique passwords (of sufficient length) for all these services without breaking a security policy. Therefore users will tend to have a small bank of passwords that are reused or simply use the same password on all systems.

Due to these weaknesses further attention was placed upon other forms of authentication. From one perspective, tokens seem to solve the underlying problem with passwords – the inability of people to remember a sufficiently long random password. By using technology, the secret knowledge could be placed in a memory chip rather than the human brain. In this fashion, the problems of needing to remember 14-character random passwords, unique to each system and regularly updated to avoid compromise, were all solved. It did, however, introduce one other significant challenge. The physical protection afforded to secret-knowledge approaches by the human brain does not exist within tokens. Theft or abuse of the physical token removes any protection it would provide. It is less likely that your brain can be abused in the same fashion (although techniques such as blackmail, torture and coercion are certainly approaches of forcefully retrieving information from people).

The key assumption with token-based authentication is that the token is in the possession of the legitimate user. Similarly, with passwords, this reliance upon the human participant in the authentication process is where the approach begins to fail. With regard to tokens, people have mixed views on their importance and protection. With house and car keys, people tend to be highly protective, with lost or stolen keys resulting in a fairly immediate replacement of the locks and appropriate notification to family members and law enforcement. This level of protection can also be seen with regard to wallets and purses – which are a store for many tokens such as credit cards. When using tokens for logical and physical access control, such as a work identity card, the level of protection diminishes. Without strong policies on the reporting of lost or stolen cards, the assumption that only the authorised user is in possession of the token is weak at best. In both of the first two examples, the consequence of misuse would have a direct financial impact on the individual, whereas the final example has (on the face of it) no direct consequence. So the individual is clearly motivated by financial considerations (not unexpectedly!). When it comes to protecting information or data, even if it belongs to them, the motivation to protect themselves is lessened, with many people unappreciative of the value of their information. An easy example here is the wealth of private information people are happy to share on social networking sites (BBC 2008a). The resultant effect of this insecurity is that tokens are rarely utilised in isolation but rather combined with a second form of authentication to provide a *two-factor* authentication. Tokens and PINs are common combinations for example credit cards.

The feasibility of tokens is also brought into question when considering their practical use. People already carry a multitude of token-based authentication credentials and the utilisation of tokens for logical authentication would only serve to increase this number. Would a different token be required to login in to the computer, on to the online bank accounts, onto Amazon and so on? Some banks in the UK have already issued a card reader that is used in conjunction with your current

Table 1.2 Biometric techniques

Physiological	Behavioural
Ear geometry	Gait recognition
Facial recognition	Handwriting recognition
Facial thermogram	Keystroke analysis
Fingerprint recognition	Mouse dynamics
Hand geometry	Signature recognition
Iris recognition	Speaker recognition
Retina recognition	
Vascular pattern recognition	

cash card to provide a unique one-time password. This password is then entered onto the system to access particular services (NatWest 2010). Therefore, in order to use the system, the user must remember to take not only the card but also the card reader with them wherever they go (or constrain their use to a single location). With multiple bank accounts from different providers this quickly becomes infeasible.

The third category of authentication, biometrics, serves to overcome the aforementioned weaknesses by removing the reliance upon the individual to either remember a password or remember to take and secure a token. Instead, the approach relies upon unique characteristics already present in the individual. Although the modern interpretation of biometrics certainly places its origins in the twentieth century, biometric techniques have been widely utilised for hundreds of years. There are paintings from prehistoric times signed by handprints and the Babylonians used fingerprints on legal documents.

The modern definition of biometrics goes further than simply referring to a unique characteristic. A widely utilised reference by the International Biometrics Group (IBG) defines biometrics as 'the automated use of physiological or behavioural characteristics to determine or verify identity' (IBG 2010a). The principle difference is in the term *automated*. Whilst many biometric characteristics may exist, they only become a biometric once the process of authentication (or strictly identification) can be achieved in an automated fashion. For example, whilst DNA is possibly one of the more unique biometric characteristics known, it currently fails to qualify as a biometric as it is not a completely automated process. However, significant research is currently being conducted to make it so. The techniques themselves can be broken down into two categories based upon whether the characteristic is a physical attribute of the person or a learnt behaviour. Table 1.2 presents a list of biometric techniques categorised by their physiological or behavioural attribute.

Fingerprint recognition is the most popular biometric technique in the market. Linked inherently to its use initially within law enforcement, Automated Fingerprint Identification Systems (AFIS) were amongst the first large-scale biometric systems. Still extensively utilised by law enforcement, fingerprint systems have also found their way into a variety of products such as laptops, mobile phones, mice and physical access controls. Hand geometry was previously a significant market player, principally in time-and-attendance systems; however, these have been surpassed by facial and vascular pattern recognition systems in terms of sales. Both of the latter techniques have increased in popularity since September 2001 for use in border control and anti-terrorism efforts. Both iris and retina recognition systems are amongst the most effective techniques in uniquely identifying a subject. Retina in particular is quite intrusive to the user as the sample capture requires close interaction between the user and capture device. Iris recognition is becoming more popular as the technology for performing authentication at a distance advances.

The behavioural approaches are generally less unique in their characteristics than their physiological counterparts; however, some have become popular due to the application within which they are used. For instance, speaker recognition (also known as voice verification) is widely utilised in telephony-based applications to verify the identity of the user. Gait recognition, the ability to identify a person by the way in which they walk, has received significant focus for use within airports, as identification is possible at a distance. Some of the less well-established techniques include keystroke analysis, which refers to the ability to verify identity based upon the typing characteristics of individuals, and mouse dynamics, verifying identity based upon mouse movements. The latter has yet to make it out of research laboratories.

The biometric definition ends with the ability to *determine* or *verify* identity. This refers to the two modes in which the biometric system can operate. To verify, or verification (also referred to as authentication), is the process of confirming that a claimed identity is the authorised user. This approach directly compares against the password model utilised on computer systems, where the user enters a username – thus claiming an identity, and then a password. The system verifies the password against the claimed identity. However, biometrics can also be used to identify, or for identification. In this mode, the user does not claim to be anybody and merely presents their biometric sample to the system. It is up to the system to determine whether the sample is an authorised sample and against which user. From a problem complexity perspective, these are two very different problems.

From a system performance perspective (ignoring the compromise due to poor selection etc.), biometric systems do differ from the other forms of authentication. With both secret-knowledge and token-based approaches, the system is able to verify the provided credential with 100% accuracy. The result of the comparison is a Boolean (true or false) decision. In biometric-based systems, whilst the end result is still normally a Boolean decision, that decision is based upon whether the sample has met (or exceeded) a particular similarity score. In a password-based approach, the system would not permit access unless all characters where identical. In a biometric-based system that comparison of similarity is not 100% – or indeed typically anywhere near 100%. This similarity score gives rise to error rates that secret-knowledge and token-based approaches do not have. The two principal error rates are:

- False acceptance rate (FAR) – the rate at which an impostor is wrongly accepted into the system
- False rejection rate (FRR) – the rate at which an authorised user is wrongly rejected from a system

Figure 1.6 illustrates the relationship between these two error rates. Mutually exclusive as neither tends towards zero, it is necessary to determine a threshold

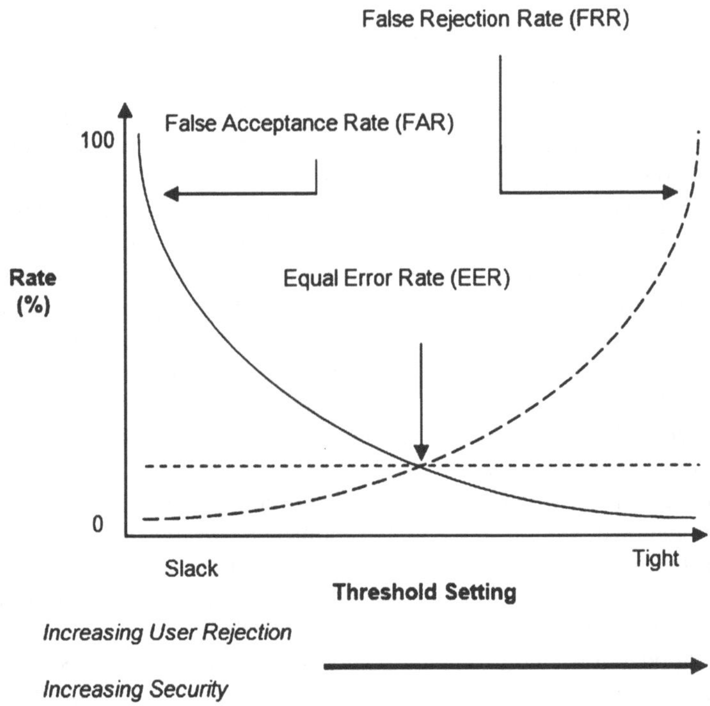

Fig. 1.6 Biometric performance characteristics

value that is a suitable compromise between the level of security required (FAR) and the level of user convenience (FRR). A third error rate, the equal error rate (EER), is a measure of where the FAR and FRR cross and is frequently used as a standard reference point to compare different biometric systems' performance (Ashbourn 2000). The performance of biometric systems has traditionally been the prohibitive factor in widespread adoption (alongside cost), with error rates too high to provide reliable and convenient authentication of the user. This has considerably changed in recent years with significant enhancements being made in pattern classification to improve performance.

Biometrics is considered to be the strongest form of authentication; however, a variety of problems exist that can reduce their effectiveness, such as defining an appropriate threshold level. They also introduce a significant level of additional work, both in the design of the system and the deployment. The evidence for this manifests itself in the fact that few off-the-shelf products exist for large-scale biometric deployment. Instead, vendors offer a bespoke design solution involving expensive consultants – highlighting the immaturity of the marketplace. Both tokens and particularly passwords are simple for software designers to implement and organisations to deploy. Further evidence of this can be found by looking at the levels of adoption over the last 10 years. Table 1.3 illustrates the level of adoption

Table 1.3 Level of adoption of authentication approaches

	2001	2002	2003	2004	2005	2006	2007	2008	2009	2010
Static account/login password	48%	44%	47%	56%	52%	46%	51%	46%	42%	43%
Smartcard and other one-time passwords	[a]	[a]	[a]	35%	42%	38%	35%	36%	33%	35%
Biometrics	9%	10%	11%	11%	15%	20%	18%	23%	26%	21%

[a]Data not available

of biometrics from 9% in 2001 rising to 21% in 2010,[1] still only a minor player in authentication versus the remaining approaches.

Whilst relatively low, it is interesting to note that adoption of biometrics has increased over the 10-year period, whilst the other approaches have stayed fairly static, if not slightly decreasing in more recent years. This growth in biometrics reflects the growing capability, increasing standardisation, increasing performance and decreasing cost of the systems.

Fundamentally, all the approaches come back to the same basic element: a unique piece of information. With something you know, the responsibility for storing that information is placed upon the user; with something you have, it is stored within the token and with something you are, it is stored within the biometric characteristic. Whilst each has its own weaknesses, it is imperative that verifying the identity of the user is completed successfully if systems and information are to remain secure.

1.4 Point-of-Entry Authentication

User authentication to systems, services or devices is performed using a single approach – point-of-entry authentication. When authenticated successfully, the user has access to the system for a period of time without having to re-authenticate, with the period of time being defined on a case-by-case basis. For some systems, a screensaver will lock the system after a few minutes of inactivity; for many Web-based systems, the default time-out on the server (which would store the authenticated credential) is 20 min and for other systems they will simply remain open for use until the user manually locks the system or logs out of the service. The point-of-entry mechanism is an intrusive interface that forces a user to authenticate. To better understand and appreciate the current use of authentication, it is relevant to examine the literature on the current use of each of the authentication categories.

[1]These figures were compiled from the Computer Security Institutes (CSI) annual Computer Crime and Abuse Survey (which until 2008 was jointly produced by the Federal Bureau of Investigation (FBI)) between the period 2001 and 2010 (CSI, 2001-2010).

Table 1.4 Top 20 most common passwords

Analysis of 34,000 passwords (Schneier 2006)		Analysis of 32 million passwords (Imperva 2010)		
Rank	Password	Rank	Password	Number of users
1	password1	1	123456	290731
2	abc123	2	12345	79078
3	myspace1	3	123456789	76790
4	password	4	Password	61958
5	blink182	5	iloveyou	51622
6	qwerty1	6	princess	35231
7	fuckyou	7	rockyou	22588
8	123abc	8	1234567	21726
9	baseball1	9	12345678	20553
10	football1	10	abc123	17542
11	123456	11	Nicole	17168
12	soccer	12	Daniel	16409
13	monkey1	13	babygirl	16094
14	liverpool1	14	monkey	15294
15	princess1	15	Jessica	15162
16	jordan23	16	Lovely	14950
17	slipknot1	17	michael	14898
18	superman1	18	Ashley	14329
19	iloveyou1	19	654321	13984
20	monkey	20	Qwerty	13856

An analysis of password use by Schneier (2006) highlighted the weakness of allowing users to select the password. The study was based upon the analysis of 34,000 accounts from a MySpace phishing attack. Sixty-five percent of passwords contain eight letters or less and the most common passwords were password1, abc123 and myspace1. As illustrated in Table 1.4, none of the top 20 most frequently used passwords contain any level of sophistication that a password cracker would find remotely challenging. Another report by Imperva (2010), some 4 years later, studied the passwords of over 32 million users of Rockyou.com after a hacker obtained access to the database and posted them online. The analysis highlighted again many of the traditional weaknesses of password-based approaches. The report found that 30% of users' passwords were six letters or fewer. Furthermore, 60% of users used a limited set of alphanumeric characters, with 50% using slang/dictionary or trivial passwords. Over 290,000 users selected 123456 as a password.

Further examination of password use reveals users are not simply content on using simple passwords but continue their bad practice. A study in 2004 found that 70% of people would reveal their computer password in exchange for a chocolate bar (BBC 2004). Thirty-four percent of respondents didn't even need to be bribed and volunteered their password. People are not even being socially engineered to reveal their passwords but are simply giving them up in return for a relatively inexpensive item. If other more sophisticated approaches like social engineering were included, a worryingly significant number of accounts could be compromised, without the need for any form of technological hacking or brute-forcing. Interestingly, 80% of those questioned were also fed up with passwords and would like a better way to login to work computer systems.

Research carried out by the author into the use of PIN on mobile phones in 2005 found that 66% of the 297 respondents utilised the PIN on their device (Clarke and Furnell 2005). In the first instance, this is rather promising – although it is worth considering the third not using a PIN represents well over a billion people. More concerning, however, was their use of the security:

- 45% of respondents never changed their PIN code from the factory default setting
- A further 42% had only changed their PIN once and
- 36% use the same PIN number for multiple services – which in all likelihood would mean they also used the number for credit and cash cards.

Further results from the survey highlight the usability issues associated with PINs that would lead to these types of result: 42% of respondents had experienced some form of problem with their PIN which required a network operator to unlock the device, and only 25% were confident in the protection the PIN would provide.

From a point-of-entry authentication perspective mobile phones pose a significantly different threat to computer systems. Mobile phones are portable in nature and lack the physical protection afforded to desktop computers. PCs reside in the home or in work, within buildings that can have locks and alarms. Mobile phones are carried around and only have the individual to rely upon to secure the device. The PIN is entered upon switch-on of the device, perhaps in the morning (although normal practice is now to leave the device on permanently), and the device remains on and accessible without re-authentication of the user for the remainder of the day. The device can be misused indefinitely to access the information stored on the device (and until reported to the network operator misused to access the Internet and make international telephone calls). A proportion of users are able to lock their device and re-enter the PIN. From the survey, however, only 18% of respondents used this functionality.

When it comes to point-of-entry authentication, misuse of secret-knowledge approaches is not unique and both tokens and biometrics also suffer from various issues. Tokens have a chequered past. If we consider their use as cash/credit cards, the level of fraud being conducted is enormous. The Association for Payment Clearing Services (APACS), now known as the UK Payments Administration (UKPA), reported the level of fraud at £535 million in 2007, a 25% rise on the previous year (APACS 2008). Whilst not all the fraud can be directly attributed to the misuse of the card, counterfeit cards and lost/stolen cards certainly can be, and they account for over £200 million of the loss. Interestingly, counterfeit fraud within the UK has dropped dramatically (71%) between 2004 and 2007, with the introduction of chip and PIN. Chip and PIN moved cards and merchants away from using the magnetic strip and a physical signature to a smartcard technology that made duplicating cards far more difficult. Unfortunately, not everywhere in the world is this new technology utilised and this gives rise to the significant level of fraud still existing for counterfeit cards.

The assumption that the authorised user is the person using the card obviously does not hold true for a large number of credit card transactions. Moreover, even with a token that you would expect users to be financially motivated to take care of, significant levels of misuse still occur. One of the fundamental issues that gave rise to counterfeit fraud of credit cards is the ease with which the magnetic-based cards could be cloned. It is the magnetic strip of the card that stores the secret information necessary to perform the transaction. A BBC report in 2003 stated that 'a fraudulent transaction takes place every 8s and cloning is the biggest type of credit card fraud' (BBC 2003). Whilst smartcard technologies have improved the situation, there is evidence that these are not impervious to attack. Researchers at Cambridge University have found a way to trick the card reader into authenticating a transaction without a valid PIN being entered (Espiner 2010).

Proximity or radio frequency identification (RFID)-based tokens have also experienced problems with regard to cloning. RFID cards are contactless cards that utilise a wireless signal to transmit the necessary authentication information. One particular type of card, NXP Semiconductor's Mifare Classic RFID card, was hacked by Dutch researchers (de Winter 2008). The hack fundamentally involves breaking the cryptographic protection, which only takes seconds to complete. The significance of this hack is in tokens that contain the Mifare chip. The chip is used not only in the Dutch transportation system but also in the US (Boston Charlie Card) and the UK (London Oyster Card) (Dayal 2008). Subsequent reports regarding the Oyster card reveal that duplicated Mifare chips can be used for free to travel on the underground (although only for a day due to the asynchronous nature of the system) (BBC 2008b). With over 34 million Oyster cards in circulation, a significant opportunity exists for misuse. Since February 2010, new cards are being distributed that no longer contain the Mifare Classic chip, but that in itself highlights another weakness of token-based approaches, the cost of reissue and replacement.

With biometric systems, duplication of the biometric sample is possible. Facial recognition systems could be fooled by a simple photocopied image of the legitimate face (Michael 2009). Fingerprint systems can also be fooled in authorising the user, using rubber or silicon impressions of the legitimate user's finger (Matsumoto et al. 2002). Unfortunately, whilst the biometric characteristics are carried around with us, they are also easily left behind. Cameras and microphones can capture our face and voice characteristics. Fingerprints are left behind on glass cups we drink from and DNA is shed from our bodies in the form of hair everywhere. There are, however, more severe consequences that can happen. In 2005, the owner of a Mercedes S-Class in Malaysia had his finger chopped off during an attack to steal his car (Kent 2005). This particular model of car required fingerprint authentication to start the car. The thieves were able to bypass the immobiliser, using the severed fingertip, to gain access. With both tokens and secret knowledge, the information could have been handed over without loss of limb. This has led more recent research to focus upon the addition of liveness detectors that are able to sense whether a real person (who is alive) is providing the biometric sample or if it is artificial.

The problem with point-of-entry authentication is that a Boolean decision is made at the point of access as to whether to permit or deny access. This decision is frequently based upon only a single decision (i.e. a password) or perhaps two with token-based approaches – but this is largely due to tokens providing no real authentication security. The point-of-entry approach provides an attacker with an opportunity to study the vulnerability of the system and to devise an appropriate mechanism to circumvent it. As it is a one-off process, no subsequent effort is required on behalf of the attacker and frequently they are able to re-access the system providing the same credential they previously compromised.

However, when looking at the available options, the approach taken with point-of-entry seems intuitively logical given the other seemingly limited choices:

- To authenticate the user every couple of minutes in order to continuously ensure the user still is the authorised user.
- To authenticate the user before accessing each individual resource (whether that be an application or file). The access control decision can therefore be more confident in the authenticity of the user at that specific point in time.

Both of the examples above would in practice be far too inconvenient to users and thus increase the likelihood that they would simply switch it off, circumvent it or maintain such a short password sequence that it was simple to enter quickly. Even if we ignore the inconvenience for a moment, these approaches still do not bypass the point-of-entry authentication approach. It still is point-of-entry, but the user has to perform the action more frequently. If an attacker has found a way to compromise the authentication credential, compromising once is no different to compromising it two, three, four or more times. So requesting additional verification of the user does not provide additional security in this case. Authenticating the user periodically with a variety of authentication techniques randomly selected would bypass the compromised credential; however, at what cost in terms of inconvenience to the user? Having to remember multiple passwords or carry several tokens for a single system would simply not be viable. Multiple biometrics would also have cost implications.

1.5 Single Sign On and Federated Authentication

As the authentication demands increase upon the user, so technologies have been developed to reduce them, and single sign on and federated authentication are two prime examples. Single sign on allows a user to utilise a single username and password to access all the resources and applications within an organisation. Operationally, this allows users to enter their credentials once and be subsequently permitted to access resources for the remainder of the session. Federated authentication extends this concept outside of the organisation to include other organisations. Obviously, for federated identity to function, organisations need to ensure relevant trust/use policies are in place beforehand. Both approaches reduce the need for the users to repeatedly enter their credentials every time they want to access a network resource.

In enterprise organisations, single sign on is also replaced with reduced sign on. Recognising that organisations place differing levels of risk on information, reduced sign on permits a company to have additional levels of authentication for information assets that need better protection. For instance, it might implement single sign on functionality with a username and password combination for all low-level information but require a biometric-based credential for access to more important data.

Both single sign on and, more recently, federated authentication have become hugely popular. It is standard practice for large organisations to implement single sign on, and OpenID, a federated identity scheme, has over a billion enabled accounts and over nine million web sites that accept it (Kissel 2009). Whilst these mechanisms do allow access to multiple systems through the use of a single credential, traditionally viewed as a bad practice, the improvement in usability for end-users has overridden this issue.

In addition to single sign on described above, there are also examples of what appear to be single sign on used frequently by users on desktop systems and browsers utilising password stores. A password store will simply store all the individual username and password combinations for all your systems/web sites. A single username and password provides access to them. This system is different from normal single sign on in that each of the resources that need access still has its own authentication credentials and the password store acts as a middle layer in providing them, assuming that the key to unlock the store is provided. In single sign on, there is only a single authentication credential and a central service is responsible for managing them. Password stores are therefore open to abuse by attacks that provide access to the local host and to the password store. A potentially more significant issue with password stores is the usability of such approaches. Whilst designed to improve usability they could in many cases inhibit use. Password stores stop users from having to enter their individual authentication credential to each service, which over time is likely to lead to users simply forgetting what they are. When users need to access the service from another computer, or from their own after a system reset, it is likely that they will encounter issues over remembering their credentials.

Single sign on and federated identity, whilst helping to remove the burden placed upon users for accessing services and applications, still only provide point-of-entry verification of a user and thus only offer a partial solution to the authentication problem.

1.6 Summary

User authentication is an essential component in any secure system. Without it, it is impossible to maintain the confidentiality, integrity and availability of systems. Unlike firewalls, antivirus and encryption, it is also one of the few security controls that all users have to interface and engage with. Both secret-knowledge and token-based approaches rely upon the user to maintain security of the system. A lost or stolen token or shared password will compromise the system. Biometrics do provide an additional level of security, but are not necessarily impervious to compromise. Current approaches to authentication are arguably therefore failing to meet the needs or expectations of users or organisations.

In order to determine what form of authentication would be appropriate, it would be prudent to investigate the nature of the problem that is trying to be solved. With what is the user trying to authenticate? How do different technologies differ in their security expectations? What threats exist and how do they impact the user? What usability considerations need to be taken? The following chapters in Part I of this book will address these issues.

References

AccessData: AccessData password recovery toolkit. AccessData. Available at: http://accessdata.com/products/forensic-investigation/decryption (2011). Accessed 10 Apr 2011

APACS: Fraud: The facts 2008. Association for payment clearing services. Available at: http://www.cardwatch.org.uk/images/uploads/publications/Fruad%20Facts%20202008_links.pdf (2008). Accessed 10 Apr 2011

Ashbourn, J.: Biometrics: Advanced Identity Verification: The Complete Guide. Springer, London (2000). ISBN 978-1852332433

BBC: Credit card cloning. BBC inside out. Available at: http://www.bbc.co.uk/insideout/east/series3/credit_card_cloning.shtml (2003). Accessed 10 Apr 2011

BBC: Passwords revealed by sweet deal. BBC News. Available at: http://news.bbc.co.uk/1/hi/technology/3639679.stm (2004). Accessed 10 Apr 2011

BBC: Personal data privacy at risk. BBC News. Available at: http://news.bbc.co.uk/1/hi/business/7256440.stm (2008a). Accessed 10 Apr 2011

BBC: Oyster card hack to be published. BBC News. Available at: http://news.bbc.co.uk/1/hi/technology/7516869.stm (2008b). Accessed 10 Apr 2011

Clarke, N.L., Furnell, S.M.: Authentication of users on mobile telephones – A survey of attitudes and opinions. Comput. Secur. **24**(7), 519–527 (2005)

Crown Copyright: Computer misuse act. Crown copyright. Available at: http://www.legislation.gov.uk/ukpga/1990/18/contents (1990). Accessed 10 Apr 2011

Crown Copyright: Data protection act 1988. Crown copyright. Available at: http://www.legislation.gov.uk/ukpga/1998/29/contents (1998). Accessed 10 Apr 2011

Crown Copyright: Regulation of investigatory powers act. Crown copyright. Available at: http://www.legislation.gov.uk/ukpga/2000/23/contents (2000a). Accessed 10 Apr 2011

Crown Copyright: Electronic communication act. Crown copyright. Available at: http://www.legislation.gov.uk/ukpga/2000/7/contents (2000b). Accessed 10 Apr 2011

Crown Copyright: Police and justice act. Crown copyright. Available at: http://www.legislation.gov.uk/ukpga/2006/48/contents (2006). Accessed 10 Apr 2011

de Winter, B.: New hack trashes London's Oyster card. Tech world. Available at: http://news.techworld.com/security/105337/new-hack-trashes-londons-oyster-card/ (2008). Accessed 10 Apr 2011

Deyal, G.: MiFare RFID crack more extensive than previously thought. Computer world. Available at: http://www.computerworld.com/s/article/9078038/MiFare_RFID_crack_more_extensive_than_previously_thought (2008). Accessed 10 Apr 2011

Espiner, T.: Chip and PIN is broken, says researchers. ZDNet UK. Available at: http://www.zdnet.co.uk/news/security-threats/2010/02/11/chip-and-pin-is-broken-say-researchers-40022674/ (2010). Accessed 3 Aug 2010

Furnell, S.M.: Computer Insecurity: Risking the System. Springer, London (2005). ISBN 978-1-85233-943-2
IBG: How is biometrics defined? International Biometrics Group. Available at: http://www.biometricgroup.com/reports/public/reports/biometric_definition.html (2010a). Accessed 10 Apr 2011
Imperva: Consumer password worst practices. Imperva Application Defense Centre. Available at: http://www.imperva.com/docs/WP_Consumer_Password_Worst_Practices.pdf (2010). Accessed 10 Apr 2011
ISO: ISO/IEC 27002:2005 information technology – Security techniques – Code of practice for information security management. International Standards Organisation. Available at: http://www.iso.org/iso/iso_catalogue/catalogue_tc/catalogue_detail.htm?csnumber=50297 (2005a). Accessed 10 Apr 2011
Kent, J.: Malaysia car thieves steal finger. BBC News. Available at: http://news.bbc.co.uk/1/hi/world/asia-pacific/4396831.stm (2005). Accessed 10 Apr 2011
Kissel, B.: OpenID 2009 year in review. OpenID Foundation. Available at: http://openid.net/2009/12/16/openid-2009-year-in-review/ (2009). Accessed 10 Apr 2011
Matsumoto, T., Matsumoto, H., Yamada, K., Hoshino, S.: Impact of artificial 'gummy' fingers on fingerprint systems. Proc. SPiE **4677**, 275–289 (2002)
Michael, S.: Facial recognition fails at Black Hat. eSecurity planet. Available at: http://www.esecurityplanet.com/trends/article.php/3805011/Facial-Recognition-Fails-at-Black-Hat.htm (2009). Accessed 10 Apr 2011
NatWest.: The secure way to get more from online banking. NatWest Bank. Available at: http://www.natwest.com/personal/online-banking/g1/banking-safely-online/card-reader.ashx (2010). Accessed 10 Apr 2011
Ophcrack.: What is ophcrack?. Sourceforge. Available at: http://ophcrack.sourceforge.net/ (2011). Accessed 10 Apr 2011
Schneier, B.: Real-world passwords. Bruce Schneier Blog. Available at: http://www.schneier.com/blog/archives/2006/12/realworld_passw.html (2006). Accessed 10 Apr 2011
Security Focus.: @Stake LC5. Security focus. Available at: http://www.securityfocus.com/tools/1005 (2010). Accessed 10 April 2011
Wood, H.: The use of passwords for controlling the access to remote computer systems and services. In: Dinardo, C.T. (ed.) Computers and Security, vol. 3, p. 137. AFIPS Press, Montvale (1977)

Chapter 2
The Evolving Technological Landscape

2.1 Introduction

Technology is closely intertwined with modern society, and few activities in our daily life do not rely upon technology in some shape or form – from boiling a kettle and making toast to washing clothes and keeping warm. The complexity of this technology is however increasing, with more intelligence and connectivity being added to a whole host of previously simple devices. For instance, home automation enables every electrical device to be independently accessed remotely, from lights and hot water to audio and visual systems. Cars now contain more computing power than the computer that guided Apollo astronauts to the moon (Physics.org 2010). With this increasing interoperability and flexibility comes a risk. What happens when hackers obtain access to your home automation system? Switch devices on, turn up the heating, or switch the fridge off? If hackers gain access to your car, would they be able to perform a denial of service attack? Could they have more underhand motives – perhaps cause an accident, stop the breaking or speed the car up? Smart electricity meters are being deployed in the US, UK and elsewhere that permit close monitoring of electricity and gas usage as part of the effort towards reducing the carbon footprint (Anderson and Fluoria 2010). The devices also allow an electricity/gas supplier to manage supplies at times of high usage, by switching electricity off to certain homes whilst maintaining supply to critical services such as hospitals. With smart meters being deployed in every home, an attack on these devices could leave millions of homes without electricity. The impact upon society and the resulting confusion and chaos that would derive is unimaginable.

With this closer integration of technology, ensuring systems remain secure has never been more imperative. However, as society and technology evolve, the problem of what and how to secure systems also changes. Through an appreciation of where technology has come from, where it is heading, the threats against it and the users who use it, it is possible to develop strategies to secure systems that are proactive in their protection rather than reactive to every small change – developing a holistic approach is key to deriving long-term security practice.

N. Clarke, *Transparent User Authentication: Biometrics, RFID and Behavioural Profiling*,
DOI 10.1007/978-0-85729-805-8_2,

2.2 Evolution of User Authentication

The need to authenticate users was identified early on in the history of computing. Whilst initially for financial reasons – early computers were prohibitively expensive and IT departments needed to ensure they charged the right departments for use – the motivations soon developed into those we recognise today. Of the authentication approaches, passwords were the only choice available in the first instance.

Initial implementations simply stored the username and password combinations in clear-text form, allowing anyone with sufficient permission (or moreover anyone who was able to obtain sufficient permission) the access to the file. Recognised as a serious weakness to security, passwords were subsequently stored in a hashed form (Morris and Thomson 1978). This provided significant strength to password security as accessing the password file no longer revealed the list of passwords. Cryptanalysis of the file is possible; however, success is largely dependent upon whether users are following good password practice or not and the strength of the hashing algorithm used.

Advice on what type of password to use has remained fairly constant, with a trend towards more complex and longer passwords as computing power and its ability to brute-force the password space improved. As dictionary-based attacks became more prevalent, the advice changed to ensure that passwords were more random in nature, utilising a mixture of characters, numerals and symbols. The advice on the length of the password has also varied depending upon its use. For instance, on Windows NT machines that utilised the LAN Manager (LM) password, the system would store the password into two separate 7-character hashes. Passwords of 9 characters would therefore have the first 7 characters in the first hash and the remaining 2 in the second hash. Cracking a 2-character hash is a trivial task and could subsequently assist in cracking the first portion of the hash. As such, advice by many IT departments of the day was to have 7- or 14-character passwords only. The policy for password length now varies considerably between professionals and the literature. General guidance suggests passwords of 9 characters or more; however, password crackers such as Ophtcrack are able to crack 14-character LAN Manager (LM) hashes (which were still utilised in Windows XP). Indeed, at the time of writing, Ophtcrack had tables that can crack NTLM hashes (used in Windows Vista) of 6 characters (utilising any combination of upper, lower, special characters and numbers) – and this will only improve in time. Unfortunately, the fundamental boundary to password length is the capacity for the user to remember it. In 2004, Bill Gates was quoted as saying 'passwords are dead' (Kotadia 2004), citing numerous weaknesses and deficiencies that password-based techniques experience.

To fill the gap created by the weaknesses in password-based approaches, several token-based technologies were developed. Most notably, the one-time password mechanism was created to combat the issue of having to remember long complex passwords. It also provided protection against replay attacks, as each password could only be utilised once. However, given the threat of lost or stolen tokens, most implementations utilise one-time passwords as a two-factor approach, combining it

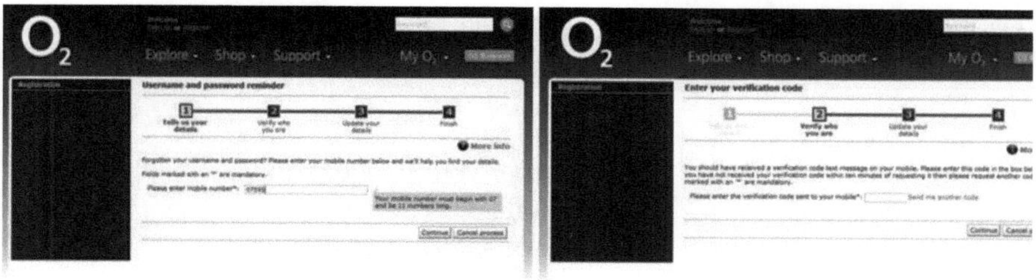

Fig. 2.1 O_2 web authentication using SMS

with the traditional username and password for instance, thereby not necessarily removing the issues associated with remembering and maintaining an appropriate password.

Until more recently, token-based approaches have been largely utilised by corporate organisations for logical access control of their computer systems, particularly for remote access where increased verification of authenticity is required. The major barrier to widespread adoption is the cost associated with the physical token itself. However, the ubiquitous nature of mobile phones has provided the platform for a new surge in token-based approaches. Approaches utilise the Short-Message-Service (SMS) (also known as the text message) to send the user a one-time password to enter onto the system. Mobile operators, such as O2 (amongst many others) in the UK, utilise this mechanism for initial registration and password renewal processes for access to your online account, as illustrated in Figs. 2.1 and 2.2.

Google has also developed its Google Authenticator, a two-step verification approach that allows a user to enter a one-time code in addition to their username and password (Google 2010). The code is delivered via a specialised application installed on the user's mobile handset, thus taking advantage of Internet-enabled devices (as illustrated in Fig. 2.3). The assumption placed on these approaches is that mobile phones are a highly personal device and as such will benefit from additional physical protection from the user than traditional tokens. The costs of deploying to these devices over providing the physical device are also significantly reduced.

The growth of the Internet has also resulted in an increased demand upon users to authenticate – everything from the obvious such as financial web sites and corporate access, to less obvious news web sites and online gaming. Indeed, the need to authenticate in many instances has less to do with security and more with possible marketing information that can be gleamed from understanding users' browsing habits. Arguably this is placing an overwhelming pressure on users to remember a large number of authentication credentials. In addition, the increase in Internet-enabled devices from mobile phones to iPads ensures users are continuously connected, consuming media from online news web sites and communicating using social networks, instant messenger and Skype. Despite the massive change in technology, both in terms of the physical form factor of the device and the increasing mobile nature of the device, and the services the technology enables, the nature of authentication utilised is overwhelmingly still a username and password.

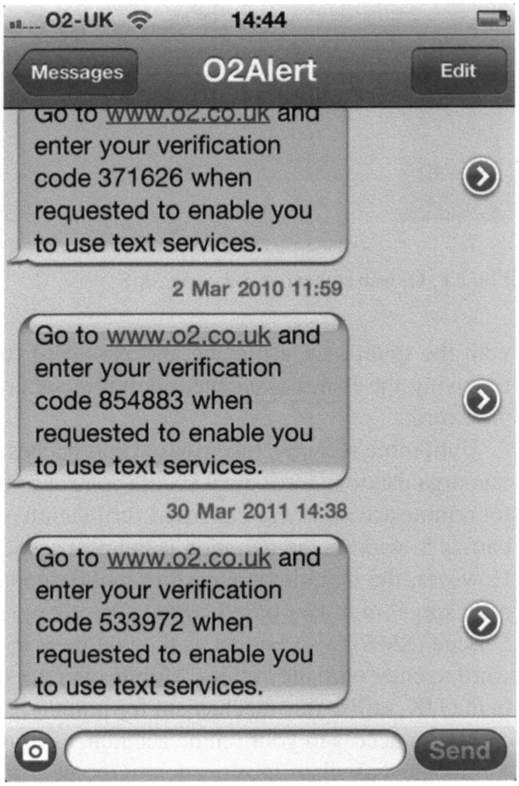

Fig. 2.2 O2 SMS one-time password

Further examination of the mobile phone reveals an interesting evolution of technology, services and authentication. The mobile phone represents a ubiquitous technology (in the developed world) with over 4.3 billion subscribers; almost two-thirds of the world population[1] (GSM Association 2010). The mobile phone, known technically as the Mobile Station (MS), consists of two components: the Mobile Equipment (ME) and a Subscriber Identification Module (SIM). The SIM is a smart card with, amongst other information, the subscriber and network authentication keys. Subscriber authentication on a mobile phone is achieved through the entry of a 4–8-digit number known as a Personal Identification Number (PIN). This point-of-entry system then gives access to the user's SIM, which will subsequently give the user network access via the International Mobile Subscriber Identifier (IMSI) and the Temporary Mobile Subscriber Identifier (TMSI), as illustrated in Fig. 2.4. Thus the user's authentication credential is used by the SIM to unlock the necessary credentials for device authentication to the network.

The SIM card is a removable token allowing in principle for a degree of personal mobility. For example, a subscriber could place their SIM card into another handset

[1] This number would include users with more than one subscription, such as a personal and business contract. So this figure would represent a slightly smaller proportion of the total population than stated.

Fig. 2.3 Google Authenticator

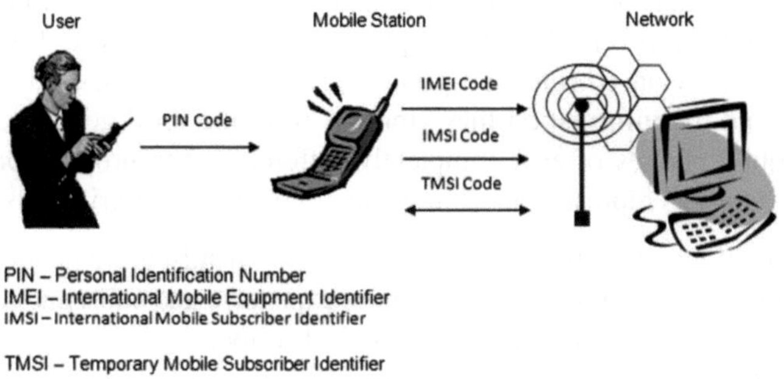

Fig. 2.4 Terminal-network security protocol

and use it in the same manner as they would use their own phone with calls being charged to their account. However, the majority of mobile handsets are typically locked to individual networks, and although the SIM card is in essence an authentication token, in practice the card remains within the mobile handset throughout the life of the handset contract – removing any additional security that might be provided by a token-based authentication technique. Indeed, the lack of using the SIM as an authenticate token has resulted in many manufacturers placing the SIM cardholder in inaccessible areas on the device, for infrequent removal.

Interestingly, the purpose of the IMSI and TMSI are to authenticate the SIM card itself on the network, and they do not ensure that the person using the phone is actually the registered subscriber. This is typically achieved at switch on using the PIN, although some manufacturers also have the PIN mechanism when you take the mobile out of a stand-by mode. As such, a weakness of the point-of-entry system is that, after the handset is switched on, the device is vulnerable to misuse should it be left unattended or stolen.

In addition to the PIN associated with the SIM card, mobile phones also have authentication mechanisms for the device itself. Whether the user is asked for the SIM-based password, the handset-based password or even both depends upon individual handsets and their configuration. The nature of the handset authentication can vary but is typically either a PIN or alphanumeric password on devices that support keyboards. Whilst the mobile phone has the opportunity to take advantage of the stronger two-factor authentication (token and password), practical use of the device on a day-to-day basis has removed the token aspect and minimised the effectiveness of the secret-knowledge approach. A survey involving 297 participants found that 85% of them left their phone on for more than 10 h a day – either switching it on at the start of the day or leaving the device switched on continuously (Clarke and Furnell 2005).

More recently, a few handset operators and manufacturers have identified the need to provide more secure authentication mechanisms. For instance, NTT DoCoMo F505i handset and Toshiba's G910 come equipped with a built-in fingerprint sensor, providing biometric authentication of the user (NTT DoCoMo 2003; Toshiba 2010). Although fingerprint technology increases the level of security available to the handset, the implementation of this mechanism has increased handset cost, and even then the technique remains point-of-entry only and intrusive to the subscriber.

More notably, however, whilst the original concept of the PIN for first-generation mobile phones may have been appropriate – given the risk associated with lost/stolen devices and the information they stored – from the third generation (3G) and beyond, mobile phones offer a completely different value proposition. It can be argued that handsets represent an even greater enticement for criminals because:

1. More technologically advanced mobile handsets – handsets are far more advanced than previous mobile phones and are more expensive and subsequently attractive to theft, resulting in a financial loss to the subscriber.
2. Availability of data services – networks provide the user with the ability to download and purchase a whole range of data services and products that can be charged to the subscriber's account. Additionally, networks can provide access to bank accounts, share trading and making micro-payments. Theft and misuse of the handset would result in financial loss for the subscriber.
3. Personal Information – handsets are able to store much more information than previous handsets. Contact lists not only include name and number but addresses, dates of birth and other personal information. Handsets may also be able to access personal medical records and home intranets, and their misuse would result in a personal and financial loss for the subscriber.

These additional threats were recognised by the architects of 3G networks. The 3GPP devised a set of standards concerning security on 3G handsets. In a document called '3G – Security Threats and Requirements' (3GPP 1999) the requirements for authentication state:

> It shall be possible for service providers to authenticate users at the start of, and during, service delivery to prevent intruders from obtaining unauthorised access to 3G services by masquerade or misuse of priorities.

The important consequence of this standard is to authenticate subscribers during service delivery, an extension of the 2G point-of-entry authentication approach, which requires continuous monitoring and authentication. However, network operators, on the whole, have done little to improve authentication security, let alone provide a mechanism for making it continuous. Even with the advent and deployment of 4G networks in several countries, the process of user authentication has remained the same.

In comparison to passwords and tokens, biometrics has quite a different history of use with its initial primary area of application within law enforcement. Sir Francis Galton undertook some of the first research into using fingerprints to uniquely identify people, but Sir Edward Henry is credited with developing that research for use within law enforcement in the 1890s – known as the Henry Classification System (IBG 2003). This initial work provided the foundation for understanding the discriminative nature of human characteristics. However, it is not until the 1960s that biometric systems, as defined by the modern definition, began to be developed. Some of the initial work was focused upon developing automated approaches to replace the paper-based fingerprint searching law enforcement agencies had to undertake.

As computing power improved throughout the 1970s significant advances in biometrics have been made, with a variety of research being published throughout this period on new biometric approaches. Early on, approaches such as speaker, face, iris and signature were all identified as techniques that would yield positive results. Whilst early systems were developed and implemented through the 1980s and 1990s, it was not until 1999 that the FBI's Integrated Automated Fingerprint Identification System (IAFIS) became operational (FBI 2011), thus illustrating that large-scale biometric systems are not simple to design and implement in practice.

With respect to its use within or by organisations, biometrics was more commonly used for physical access control rather than logical in the first instance. Hand geometry found early applications in time and attendance systems. The marketplace was also dominated with vendors providing bespoke solutions to clients. It simply wasn't possible to purchase off-the-shelf enterprise solutions for biometrics; they had to be individually designed. Only 9% of respondents from the 2001 Computer Crime and Abuse Survey had implemented biometric systems (Power 2001). Significant advances have been made in the last 10 years with the development of interoperability standards to enable a move away from dedicated bespoke systems to providing choice and a flexible upgrade path for customers, these efforts demonstrating the increasing maturity of the domain.

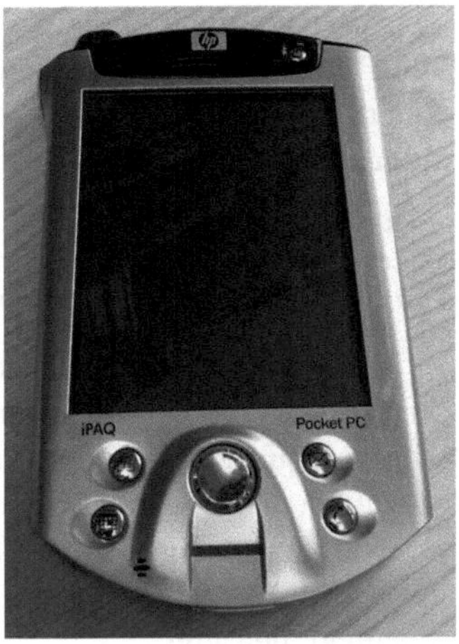

Fig. 2.5 HP iPaq H5550 with fingerprint recognition

Biometrics is now used in a variety of industries including national/government, law enforcement, transportation, military, retail, financial services, health care and gaming (IBG 2010c). Indeed, an increasing variety of personal end-user technologies, such as mobile phones, laptops and even cars have biometric technologies incorporated within them. An early implementation of fingerprint recognition on a Personal Desktop Assistant (PDA) was the Hewlett-Packard iPaq H5550, illustrated in Fig. 2.5.

The nature of authentication has certainly evolved since the early days of computing, with many more options now available to organisations that go beyond the simple password. However, whilst incredible enhancements have taken place with the underlying technology, operating systems and applications over the past 30 years, the most common form of authentication has remained identical. Indeed, even when standards describe the need for change and technology evangelists warn of the risks of still using passwords, passwords are still almost ubiquitously implemented. Whilst there are many reasons for this (and this book will investigate many of them), the lack of a systematic advancement in authentication technologies versus the systems they support is a significant and frequently unrecognised risk to the underlying information.

2.3 Cyber Security

The nature and scale of the security threat and those motivated to take advantage of them informs what protection is required. It is therefore prudent to appreciate the severity of cybercrime to date, and identify trends that currently exist and future directions of concern. To aid in this understanding, the security domain produces a variety of cybercrime and security-related surveys. Each of the surveys has a different focus; however, their purpose collectively is to provide an understanding of the cybercrime landscape and what organisations are currently doing to tackle or more specifically mitigate the problems. Whilst many surveys exist, their methodologies and statistical relevance vary considerably, and five in particular are regularly cited in literature. They are also published annually, which provides a rich source of information to better appreciate the changing nature of the threats. The surveys of particular interest are:

- Computer Security Institute's (CSI) Computer Crime and Security Survey (Richardson 2009). Amongst the most popular to reference, this survey's focus is upon understanding what threats organisations are facing and what measures they are putting in place to mitigate them. The 2009 edition consisted of 443 participants from a wide range of sectors in the US, including government and academia. The nature of the participants means the survey is skewed towards enterprise-level organisations.
- PriceWaterHouseCoopers Global State of Information Security Survey (PWC 2010a). This survey is a more business rather than technically focused survey looking at the effects information security has on the business. The 2011 survey had over 12,000 participants from 135 countries and represents a far more global picture of opinion on the information security domain.
- Ernst and Young's Global Information Security Survey (Ernst and Young 2010). Another global information security survey. Whilst smaller in the number of participants, with the 12th annual report consisting of 1,900 participants, this survey includes a number of pertinent questions and focuses more on the technical aspects of information security.
- InfoSecurty Europe's Information Security Breaches Survey (formally undertaken by the Department for Business, Enterprise and Regulatory Reform (BERR)) (PWC 2010b). This is a UK-based survey rather than US or global and provides a useful insight into the issues related to UK organisations. It has a skew towards Small-to-Medium Enterprises (SMEs), which enables a useful contrast against the CSI survey to identify differences in experiences and priorities. The 2010 survey had 539 responses.
- Symantec's Global Internet Security Threat Report (Symantec 2010). This report differs from the others in that it does not rely on people to respond but rather utilises a number of technical measurements. This removes any subjectivity from the response and provides a snapshot of the cybercrime problem. It does this by taking various measurements, including utilising more than 240,000 sensors across 200 countries. The downside of this report is that, whilst providing the most accurate perspective of the problem, it lacks an appreciation and coverage of what organisations are doing in response.

The motivations of attackers have varied considerably: from the preserve of the technically elite looking to demonstrate their technical prowess to an era of script kiddies 'playing' with technologies – causing significant problems to information systems but lacking any real knowledge of how or what they were doing. Today, cybercrime has taken a far more sinister turn. Motivations have predominately become financially motivated and computer misuse is becoming increasingly prevalent amongst all aspects of society. The reasons for this are many, but the underlying increase and reliance upon computers and the Internet have provided a wider opportunity for misuse.

From an analysis of the literature, several (interrelated) areas of cybercrime and security appear particularly relevant to discuss:

- Malicious software (malware) and botnets
- Social engineering
- Financial fraud
- Technical countermeasures

Malware, considered the mainstay of cybercrime, is a category of misuse that includes virus, worms and Trojans. The Elk Cloner virus in 1981 is regarded as the first virus discovered in the wild. Since then, a large number of viruses have been created, with varying levels of severity from benign annoyance to severely crippling systems. Over time, the focus has, however, evolved away from viruses towards worms in the late 1990s and more recently onto Trojans. This transition in the use of malware mirrors the evolution of the uses of the computer and their applications. Viruses were popular for demonstrating what is possible, from crashing systems to being a simple irritation. Whilst not appreciated at the time, they were however limited in that they relied upon human intervention to propagate between systems. Whilst there was no shortage of unwitting volunteers, the scope for widespread global infection was restricted. With the birth of the Internet and global networking came an opportunity to exploit this in the connected world. Worms have the ability to self-propagate, which enabled them to spread globally with little to no user input at all. Worms propagate by having the ability to take advantage of vulnerabilities that exist in OSs or applications. If the vulnerability is not patched, the worm is able to exploit it. For instance, the Slammer (also known as the Sapphire) worm in 2003 spread to 750,000 hosts globally in just 31 min (Moore et al. 2003). In fact, most of these hosts were infected in less than 10 min. More recently Trojans have been favoured. Unlike the traditional virus or worm that has an initial big impact on the system – whether it is by crashing the system or using it to self-propagate, malicious code writers have identified far more financially rewarding directions for their efforts. Trojans are designed to remain hidden as much as possible, typically sniffing sensitive information such as usernames, passwords and more commonly bank account details. An analysis of the 2009 Global Internet Threat Report (Symantec 2010) highlights these issues:

- 5,724,106 malicious code signatures exist
- Of the top ten malicious software, six were Trojans, three worms and one virus
- Four out of the ten most prevalent malicious threats exposed confidential information

Interestingly, Symantec notes that of the 5.7 million code signatures it has, over 51% of them were created in 2009 alone, demonstrating a significant increase in the prevalence of new malware. These trends are supported by the Computer Crime and Abuse Survey, which in the same year reported the following (Richardson 2009):

- Respondents indicated malware infection increased with 64% of respondents registering an incident
- Incidents of financial fraud increased to 19.5% of respondents
- Password sniffing increased to 17.3% from 9% the previous year
- Denial of service attacks increased to 29.2%

The nature of the attacks were also becoming more focused, with the CSI survey reporting 24.6% of the incidents involved targeted attacks against organisations. This highlights a growing level of sophistication being deployed by the hacker in order to compromise systems. It was noted by the Symantec report that Advanced Persistent Threat (APT) attacks are specifically designed to go undetected for long periods of time in order to gather information – with the subsequent consequence that large numbers of identities are compromised.

The widespread nature of malware is also highlighted by the impact of botnets. Botnets are networks of zombie computers under the control of a malicious user. These networks can be used to store illegal materials, distribute spam email or be used in *Distributed Denial of Service* (*DDoS*) attacks. Symantec noted over 6.7 million distinct bot-infected computers, with over 46,000 active bot-infected computers per day (on average) (Symantec 2010). Whilst having no direct effect upon authentication credentials, the fact the software is able to take control of the computer suggests that it has sufficient permission to perform whatever it would want to, and with over 6.7 million hosts infected, that represents a significant number of insufficiently protected systems. The Symantec report also notes that these bot-networks are available for purchase with prices in the area of $0.03 per bot. Given these bot-networks can be on the order of tens of thousands in size, this soon becomes a significant revenue source.

The desire to obtain identities or access credentials is also experienced by other threats. The most significant threat in this direction is from Phishing – unsolicited emails that purport to come from a legitimate organisation asking a user to reaffirm their authentication credentials. The Anti-Phishing Working Group reported for the first quarter in 2010 just over 30,000 unique Phishing messages (down from an all-time high of 56,000 in August 2009) (APWG 2010b). This social engineering approach seeks to take advantage of people's lack of security awareness, naivety and willingness to comply. Whilst Phishing messages were originally presented in a rather unprofessional manner in terms of the language used and the layout, making them easier to identify, more recent messages are very professional, with few indicators to tell them apart from legitimate messages (as illustrated in Fig. 2.6).

Social engineering attacks such as Phishing do represent a move away from traditional social engineering attacks that seek to obtain authentication credentials of systems rather than specific applications. Unsurprisingly this is linked to the desire for financial reward, with easier gains to be obtained through compromising bank accounts rather than computers. It is difficult to ascertain the scale of this attack in financial terms, but credit-card fraud alone cost the UK £440 million in 2009 and Internet-based banking fraud cost £59 million (Financial Fraud UK 2010). Whilst

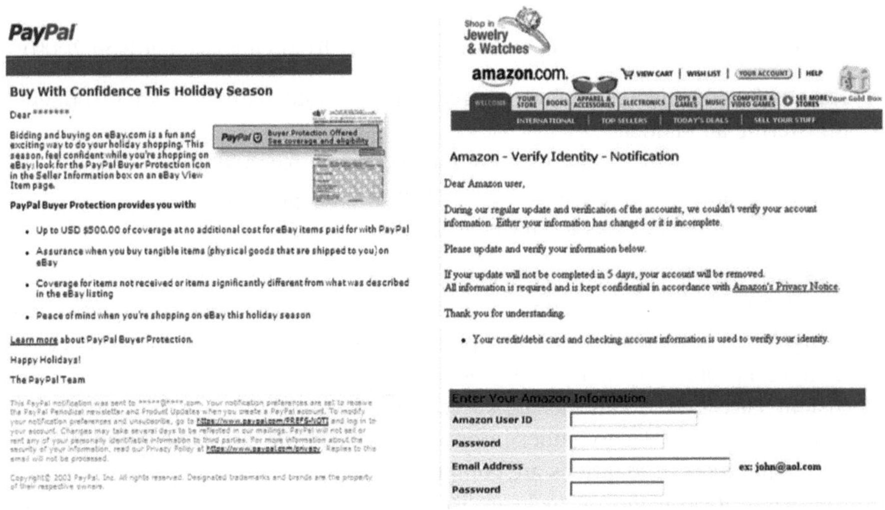

Fig. 2.6 Examples of phishing messages

there are a variety of attacks that would have led to this loss, increasingly social engineering attacks such as Phishing comprise an increasingly sizeable share.

Whilst significant focus has been placed on the compromise of credentials upon individual systems, hackers have also sought to compromise organisational databases that also store such information. For online banking, the user's credentials need to be verified against what the bank holds for that particular user and this is the same process for any online service where the user needs to log in. Whilst typically representing a more challenging attack, as organisations are generally more proficient at protecting systems than home users, this is not always the case. In practice financial institutions do tend to be far better protected, as one might expect. However, there are many other organisations that deal with our information which do not necessarily take the same steps to protect our information. A successful attack also has the advantage of compromising many users' accounts rather than just one. The most notable example to date is TK Maxx, where over 46 million payment cards were compromised with a potential cost to the company of £800 million (Dunn 2007). The UK Government also has a rather poor record of maintaining its citizens' data:

- In June 2008, HM Revenue and Customs lost discs containing personal information on 25 million citizens – potentially worth up to £1.5 billion to criminals (BBC 2007).
- A loss of a Ministry of Defence (MoD) laptop revealed passport, national insurance and driving license information on 173,000 people, which also included banking details for over 3,700 individuals (BBC 2008a).
- Six laptops stolen from a hospital with over 20,000 patient details (BBC 2008b)

Whilst the above examples were not results of hacking but rather simple neglect by the data owners, they do present two key concerns:

1. Individual user accounts and private information are being compromised.
2. The security being deployed to protect these systems is not effective at protecting information.

From an authentication perspective these compromises raise a number of issues. In some cases they have direct access to authentication credentials to access accounts or services. They also have enough personal information to be able to create accounts and masquerade as the legitimate user. Lastly, they could utilise this personal information to attempt to compromise passwords on other accounts and systems – as many people will base their secret knowledge on their personal information. Determining the correct identity of the user and mitigating against targeted misuse is therefore paramount for systems and services to remain secure.

When looking at what organisations are utilising in terms of technology, the CSI survey notes 46% of its respondents utilised account logins/passwords, 36% smartcards and other one-time tokens and 26% biometrics (Richardson 2009). Whilst this is not unexpected given the previous commentary, the report goes further by asking how satisfied respondents are with the technologies. Interestingly, none of the 22 categories of security technology scored particularly high on the level of satisfaction, but both passwords and biometrics score a 3 (on a scale of 1–5), with tokens scoring just over 3.5. This highlights a somewhat systemic dissatisfaction with security technologies, where none of the categories scored above 4, and raises a question as to what degree are security technologies fit for purpose – where the purpose is being defined by those who have to implement, configure and maintain systems in practice. When respondents were asked which security technology they would wish for, second in the list went to identity and access management technologies. Specific comments made included 'cheap but effective strong authentication for web sites that is not subject to man-in-the-middle attacks', 'a single sign-on product that works', 'mandatory biometrics use', 'something to completely replace the common password' and 'better multi-factor authentication solutions'. These comments demonstrate a huge desire for more effective means of managing identity verification, highlighting that few current products or services are available to solve the problem.

The final aspect of this cyber security section is focused upon a more recent evolution of malicious software that has given rise to a significant new breed of threat. This trend is a transition of malware affecting the cyber space into the physical world, having physical repercussions that potentially could affect life and limb. Whilst such malware has been theoretically discussed for many years, few real-life examples have ever existed. This changed in 2010 with the introduction of the Stuxnet worm (Halliday 2010). It was specifically designed to attack Supervisory Control and Data Acquisition (SCADA) systems and included the capability to reprogramme programmable logic controllers (PLCs). In particular, this worm focused upon a Siemens system, which is reported to be used in a number of high-value infrastructures, including Iran's Bushehr Nuclear facility.

This new breed of malware will inevitably lead to more approaches that seek to take advantage of commercial systems previously thought to be safe. With ever-increasing connectivity, such as electricity smart meters in every home and cars that can be remotely diagnosed, the possibility for misuse of these systems is only going to increase. Whilst malware frequently utilises vulnerabilities to compromise systems to propagate and compromise the system rather than comprising authentication credentials, it is clear that as authentication is the first barrier of defence, inappropriate use of authentication would only serve to simplify the process of compromise on these future systems which could lead to significant personal harm.

2.4 Human Aspects of Information Security

The human aspects of information security are becoming increasingly recognised as an important aspect of security countermeasures that require any form of interaction from a user – whether it be installation, configuration, management or response. The phrase 'security is only as strong as the weakest link' highlights that too often the weakness turns out to be human-related. Whether it is social engineering, an inability to use security countermeasures, the complexities of configuration information, poor awareness and knowledge of threats or merely disenfranchised users fed up of security controls preventing them from undertaking the tasks they need to, users inevitably are the single largest cause of security breaches.

It is not always the end-user's fault. As the sophistication of threats increase, a corresponding increase can also be identified in relying upon users to make appropriate decisions. The number of systems, applications and services are all significantly increasing in number and complexity. This can merely serve to confuse and disorientate users. Furthermore, what is required on one platform can differ on another. For instance, the more familiar PC would typically have Anti-Virus, a Firewall and Anti-Spyware security controls installed almost by standard. A user would be forgiven for thinking such technologies are also appropriate for any other technologies that connect to the Internet such as the mobile phone, PDA, tablets or games consoles. However, this is not the case, with each technology platform requiring a different set of controls. Given these technologies are no longer the preserve of the technological elite, is it sensible to assume that users have sufficient detailed knowledge of the threats and subsequent security controls required for each individual platform they might utilise? Even when software by the same company is developed for different technology platforms, users can expect to have very different experiences. For instance, under Microsoft Windows XP, users can expect to have a more unified experience of managing security through the Security Centre. However, Microsoft's Mobile software has a distributed approach to security, with features in various places throughout the system (Botha et al. 2009). A user well informed on the use of Windows XP systems would subsequently need to learn a whole new way of protecting his or her mobile device, even though both systems are developed by the same company.

From a human interface perspective, authentication technologies are one of the few security controls that users have no choice in utilising. Issues resulting from poor usability of authentication technologies have already been identified in terms of poor password use and the reuse of passwords on multiple systems to name a few. Token-based approaches certainly have fewer issues fundamentally but also provide a lower level of security and as such are typically implemented with two-factor

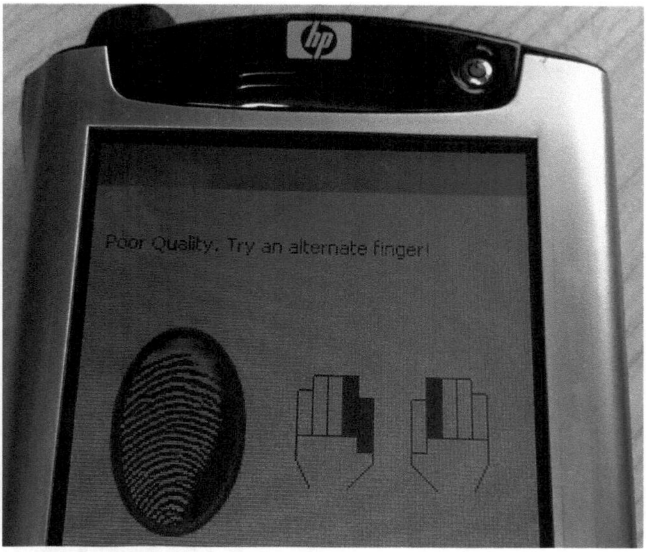

Fig. 2.7 Fingerprint recognition on HP PDA

authentication alongside PINs. Biometrics also suffers from a variety of usability problems, from the inability to successfully capture biometric samples to the incorrect rejection of the authorised user. Figure 2.7 illustrates the problems of fingerprint acquisition on a PDA. If the user swipes too quickly, too slowly or at an incorrect angle, the system is unable to obtain an image.

Arguably, certainly authentication technologies are more prone to human-based threats than others, taking social engineering as an example. Secret-knowledge and to a certain degree token-based approaches provide an opportunity for the user to reveal or release the authentication credential to the attacker. For example, the 2004 study found 70% of people would reveal their computer-based password in exchange for a chocolate bar (BBC 2004) or skimming of credit cards (BBC 2003). Inherently with biometric-based techniques, the user has no direct knowledge of the information and therefore is not subjective to social engineering in the same fashion. Moreover, people appear more reluctant to provide such information in the first place. Biometric techniques do, however, introduce another issue of latent samples; with users potentially leaving biometric samples all over the place – latent fingerprints on drinking glasses – facial samples can be easily obtained via widely available digital cameras and voice samples recovered from microphones. That said, significant research effort is now being placed on the liveness of samples to ensure they are fresh and from legitimate individuals.

System designers have also attempted to improve the situation, with password-based mechanisms including password reminders or mechanisms for recovering or resetting a forgotten password. The degree to which these are seen as useful is questionable, with many systems relying upon further secret or cognitive-based knowledge in order to reset the password – further exacerbating the need for users

Fig. 2.8 UPEK Eikon fingerprint sensor

to remember the appropriate responses. In addition, the use of cognitive-based questioning itself can introduce security weaknesses, with many questions resulting in a limited number of possible results, thereby effectively reducing the password space. Considerable effort has also been expended on developing biometric systems that are more effective in capturing biometric samples, either through technology means by modifying the captured sample in a usable form or by developing a capture system that ensures that the user provides a usable sample through presenting his or her features in the correct fashion. Figure 2.8 illustrates a swipe-based fingerprint sensor that has attempted to resolve the issues of orientation by including a physical groove with the sensor to guide the finger. Notably, this still does not resolve issues of speed of swiping.

As it is unlikely that the threat landscape will subside, the need for systems to rely upon users to make (the right) decisions will remain. Traditionally, little resource has been given to the niceties of good human–computer design. Effort has been focused on ensuring the software is able to achieve its aims and if appropriate improve the performance by which it does so. It is therefore imperative for system designers to give more careful thought to the design and operation of their security systems. Arguably with point-of-entry authentication technologies, a need will always exist for human–computer interaction and whilst developments can be made both on the form factor of the capture device and the interfaces, developing a single technology to service the needs of all users (with all their differing technological backgrounds and security requirements) will remain an incredibly challenging task. It is therefore necessary to seek approaches that either incorporate better human–computer interaction or remove the necessity for the user to need to interact.

2.5 Summary

The rapidly evolving technological and threat landscape places a real and considerable burden upon individuals. Whilst security awareness initiatives can seek to improve people's understanding of the problems associated with using technology and in particular the Internet, the speed of change in threats will perpetually leave a gap between users' knowledge and the threats looking to exploit users.

Technology-based threats will continue to attack systems and new strains of malware will find approaches to circumvent security countermeasures. In the end, however, technology-based countermeasures will be devised to mitigate and reduce the risk of attack. Attackers will therefore continue to focus upon end-users themselves – taking advantage of their lack of awareness (or perhaps greed) to circumvent security controls and enable misuse of systems or information. Better awareness, better design of security interfaces and functionality will enable users to be better placed to make the correct decision.

Whilst individual authentication mechanisms deployed on systems and services might meet security requirements, providers fail to acknowledge the inconvenience they place upon users given their need to access not just one system or service but a great many. With such a burden on users to remember credentials for so many systems and applications, users have no choice but to fall into bad practices, using the same credentials for multiple systems and relying on weaker passwords. Procedures have been developed to enable users having difficulty in accessing systems; however, they are further inconveniencing the user, forcing them to complete a further set of questions and on occasions reducing the effective level of security. This problem is further exacerbated by the need for so many services requiring the user to authenticate, when in actuality little need for authentication of the user is required.

Therefore, in order to ensure authentication is secure, reliable and convenient, it is prudent to investigate the actual need for authentication in various contexts and analyse how this could be best achieved. The following chapter describes how authentication actually works and highlights the huge disparities between expectations of security and the reality of operations.

References

3GPP: 3G security: security threats and requirements. 3G partnership project. Technical specification group services and system aspects. Document 3G TS 21.133 version 3.1.0 (1999)

Anderson, R., Fuloria, S.: Who controls the off switch. Cambridge University. Available at: http://www.cl.cam.ac.uk/~rja14/ (2010). Accessed 10 Aug 2010

APWG: Phishing activity trends report: 1st quarter 2010. Anti-Phishing Working Group. Available at: http://www.antiphishing.org/reports/apwg_report_Q1_2010.pdf (2010b). Accessed 10 Apr 2011

BBC: Credit card cloning. BBC Inside Out. Available at: http://www.bbc.co.uk/insideout/east/series3/credit_card_cloning.shtml (2003). Accessed 10 Apr 2011

BBC: Passwords revealed by sweet deal. BBC News. Available at: http://news.bbc.co.uk/1/hi/technology/3639679.stm (2004). Accessed 10 Apr 2011

BBC: Data lost by revenue and customs. BBC News. Available at: http://news.bbc.co.uk/1/hi/7103911.stm (2007). Accessed 10 Apr 2011

BBC: More MoD laptop thefts revealed. BBC News. Available at: http://news.bbc.co.uk/1/hi/7199658.stm (2008a). Accessed 10 Apr 2011

BBC: Six laptops stolen from hospital. BBC News. Available at: http://news.bbc.co.uk/1/hi/7461619.stm (2008b). Accessed 10 Apr 2011

Botha, R.A., Furnell, S.M., Clarke, N.L.: From desktop to mobile: examining the security experience. Comput Secur **28**(3–4), 130–137 (2009)

Clarke, N.L., Furnell, S.M.: Authentication of users on mobile telephones – a survey of attitudes and opinions. Comput Secur **24**(7), 519–527 (2005)

Dunn, J: Police breakthrough on stolen TK Maxx Data. IDG Inc. Available at: http://www.computerworlduk.com/news/security/4789/police-breakthrough-on-stolen-tk-maxx-data/#%23 (2007). Accessed 10 Apr 2011

Ernst and Young: 13th global information security survey 2010. Ernst & young. Available at: http://www.ey.com/GL/en/Services/Advisory/IT-Risk-and-Assurance/13th-Global-Information-Security-Survey-2010—Information-technology–friend-or-foe- (2010). Accessed 10 Apr 2011

FBI: Integrated automated fingerprint identification system (IAFIS). Federal bureau of investigation. Available at: http://www.fbi.gov/about-us/cjis/fingerprints_biometrics/iafis/iafis (2011). Accessed 10 Apr 2011

Financial Fraud UK. Fraud: the facts 2010. Financial fraud UK. Available at: http://www.cardwatch.org.uk/images/uploads/publications/Fraud%20The%20Facts%202010.pdf (2010). Accessed 9 Nov 2010

Google: Google authenticator. Google. Available at: http://www.google.com/support/a/bin/answer.py?hlrm=en&answer=1037451 (2010). Accessed 10 Apr 2011

GSM Association: Mobile phone statistics. GSM association. Available at: http://www.gsmworld.com/newsroom/market-data/market_data_summary.htm (2010). Accessed 10 Apr 2011

Halliday, J: Stuxnet worm is the 'work of a national government agency'. Guardian News and Media. Available at http://www.guardian.co.uk/technology/2010/sep/24/stuxnet-worm-national-agency (2010). Accessed 10 April 2011

IBG: The Henry classification system. International biometric group. Available at: http://static.ibgweb.com/Henry%20Fingerprint%20Classification.pdf (2003). Accessed 10 Apr 2011

IBG: Biometrics market and industry report 2009–2014. International Biometric Group. Available at: http://www.biometricgroup.com/reports/public/market_report.php (2010c). Accessed 10 Apr 2011

Kotadia, M: Gates: the password is dead. Long live the SecurID. Silicon.com. Available at: http://www.silicon.com/technology/security/2004/02/26/gates-the-password-is-dead-39118663/ (2004). Accessed 10 Apr 2011

Moore, D., Paxson, V., Savage, S., Shannon, C., Staniford, S., Waever, N.: Inside the slammer worm. IEEE Secur Priv **1**(4), 33–39 (2003)

Morris, R., Thompson, K: Password security: a case history. Bell Laboratories. Available at: http://cm.bell-labs.com/cm/cs/who/dmr/passwd.ps (1978). Accessed 10 Apr 2011

DoCoMo: NTT DoCoMo unveils ultimate 3G i-mode phones: FOMA 900i series. NTT DoCoMo. Available at: http://www.nttdocomo.com/pr/2003/001130.html (2003). Accessed 10 Apr 2011

Physics.org: Your car has more computer power than the system that guided Apollo astronauts to the moon. Institute of Physics. Available at: http://www.physics.org/facts/apollo.asp (2010). Accessed 10 Apr 2011

Power, R.: CSI/FBI computer crime and security survey. Computer Security Issues & Trends, Computer Security Institute **7**(1) (2001)

PWC. Global state of information security survey. PricewaterhouseCoopers. Available at: http://www.pwc.com/gx/en/information-security-survey (2010a). Accessed 10 Apr 2011

PWC: Information security breaches survey. Infosecurity Europe. Available at: http://www.pwc.co.uk/eng/publications/isbs_survey_2010.html (2010b). Accessed 10 Apr 2011

Richardson, R: CSI computer crime and security survey. Computer Security Institute. Available at: www.gocsi.com (2009)

Symantec: Internet security threat report. Symantec. Available at: http://www.symantec.com/business/threatreport/index.jsp (2010). Accessed 10 Apr 2011

Toshiba: Toshiba Portege G910. Toshiba. Available at: http://www.toshibamobile-europe.com/Products/G910/Technicalspecs.aspx (2010). Accessed 10 Apr 2011

Chapter 3
What Is Really Being Achieved with User Authentication?

3.1 Introduction

The term identity verification is frequently used to describe a desire to secure access to systems or services, whether that be to access a computer, mobile phone, cash machine (automated teller machine (ATM)), Amazon account, or government services, purchase movies through iTunes or travel internationally. Whilst arguably a need exists in all of these particular examples, a question that arises is whether the current mechanisms in place to provide identity verification are actually achieving the level of security required.

In addition, whilst one may wish to authenticate an individual with 100% accuracy, realistically this is not possible. Secret-knowledge and token approaches might technically provide a Boolean decision; however, they rely upon a number of assumptions that invariably are false or questionable. Similarly, biometric approaches are simply unable to achieve 100% accuracy. It is therefore imperative that due consideration be given to the fallibilities of authentication technologies and an appropriate solution be put in place to actually meet the level of security required. Moreover, given that the level of security in the identity of the user is likely to differ depending upon the application, it is clear that one-size-fits-all approaches are unsuitable and thus approaches should be tailored to meet the specific risks of the applications or systems. For instance, the mobile phone is protected through the use of personal identification number (PIN). Once authenticated, a user is able to send a simple, relatively cheap, text message or make a 4 h-long international call. From a security perspective, both operations have significantly different impacts. Is it fair to burden the user with intrusive and inconvenient authentication for operations that carry little financial or information risk? Arguably, point-of-entry authentication mechanisms are valuable for ensuring correct access to the system initially, but fail to subsequently verify the user when actually undertaking operations that have significant financial and personal/corporate risks associated to them.

Through understanding why authentication is needed and what is being achieved, a better understanding of the underlying requirements can be developed that will enable appropriate technologies to be deployed.

N. Clarke, *Transparent User Authentication: Biometrics, RFID and Behavioural Profiling*,
DOI 10.1007/978-0-85729-805-8_3,

3.2 The Authentication Process

Authentication for whatever purpose has been motivated by a need to ensure the legitimate individual is able to access their resources. Locks were developed to ensure that thieves do not have easy access to homes; signatures and banking passbooks existed to ensure an individual could access their money and passports enable successful border control. Whilst some might argue about the relevant merits of the authentication techniques involved, all three examples given are suitably addressed through point-of-entry authentication. There is little to be gained in extending the authentication process beyond point-of-entry. For instance:

- Having gained access to a house, valuable items around the house can be stolen in a relatively short time frame. Albeit, one could place smaller more valuable objects in a secondary container such as a safe.
- Once authenticated at a border crossing, it would be infeasible to re-authenticate the individual unless he could be located within the population.
- Once access to a bank account has been provided, the individual only requires a short window of opportunity to clear any money from the account. Any re-authentication of the user would simply be too late.

Whilst all the previous examples are still valid solutions to the authentication problem, the nature of the problem has arguably evolved to encompass a new and far wider set of risks, particularly as technology and individuals' use of it has changed. With this new set of risks, one would expect a revised approach to solving the authentication issue. However, all too often, usability, time to market and expense come at the cost of security. To illustrate the point, a series of examples are presented below:

- ATM debit cards – a two-factor authentication approach utilising the card as a token and a 4-digit PIN or physical signature. The PIN is used when obtaining money from a cash machine and the signature for verification when purchasing goods in a shop. A more recent transition in the UK (and wider EU) is to remove the signature for shop purchasing in favour of the PIN – referred to as Chip and PIN. Chip and PIN is now universally used throughout by merchants. Whilst the physical signature approach was certainly an ineffective measure, widening the scope of the PIN has also increased the opportunity for misuse. Under the old model, it was relatively difficult for attackers to misuse the debit card in order to obtain cash. Threats against accessing the card itself have remained, in that it can be stolen. But under the previous model, the risks of shoulder surfing were minimised because entry of the PIN *only* occurred at cash machines. The same cash machines included additional security countermeasures to hide PIN entry and (potentially) photograph users. Under the new model, individuals now have to enter the PIN each and every time they use it in a shop – using handsets that are far easier to shoulder surf. Arguably, debit cards would benefit from having differing point-of-entry authentication approaches applied depending upon the context.

- E-Commerce web sites – these provide an electronic store for many of the credit and debit cards users have. They protect access to this information through a standard password-based approach. Traditionally little protection was provided on ensuring a significantly complex password, so accounts had little protection from misuse. More common (to larger online companies) is to provide a password complexity indicator when creating the account. This provides an indication as to the strength of the password being created. However, too frequently these indicators do not really provide an effective measure, with relatively simple name and two-digit (e.g. luke23) combinations measuring as strong on the indicator. Due to levels of credit card fraud, companies have introduced a secondary measure when purchasing goods or services on the card – a further password-based approach that is specifically linked to the card the person is using (rather than the e-commerce company account). So the solution to e-commerce fraud, which utilises a password-based mechanism to protect the account, is to add another password-based approach. Moreover, the setting up and bypassing of this secondary measure can be completed with information from the credit card and knowledge of the postcode of the house the card is registered with. Interestingly, banks are beginning to understand the risks posed by Internet-based operations and a number of them have rolled out multi-factor authentication involving card readers and one-time passwords.
- Network protocols – these provide the ability for users to get access to systems and information. Whilst most users will be unaware of what protocols are, many will use them regularly to gain access to systems. Unfortunately, some of the protocols communicate with each other and send authentication data in cleartext (i.e. in human readable form). They do not perform any encryption or protection of the data. The most popular is possibly Post Office Protocol 3 (POP3), which is used to receive email from your server, but others exist, such as Telnet (regularly used to connect to other computers). Any computer connected to an unsecured wireless network would be able to sniff the network traffic undetected and obtain the authentication credentials (as well as all email and other data sent using the protocol). Whilst mechanisms do exist for securing these protocols, many systems still use the unsecured default settings. This example also demonstrates the designers' lack of consideration to security. It is unlikely that no one but the technically savvy user would appreciate what the underlying protocol is doing with his or her information. Should users decide to use the same password on multiple systems/applications, this could enable significantly larger volumes of data compromise.

The previous examples have helped to demonstrate that too frequently authentication is not necessarily given due consideration or due priority. Indeed, it often appears that authentication mechanisms are present merely to facilitate the delivery of services rather than to provide any true security. For instance, what is the purpose of having a login on an e-commerce web site?

- Is it to provide security?
- Is it to enable quick purchase of goods through storing credit card information?
- Is it to provide the e-commerce web site information about what you are looking at and what you are buying?
- Is to provide an after-sales service to enable tracking and returning of products?

Of these possible responses, providing security is probably of a lower priority. It certainly is the only response that does not have a direct impact upon revenue (although it certainly will have an indirect effect). In other scenarios, it appears that authentication is merely an afterthought. An initial implementation of Chip and PIN enabled some people to pay for goods in a supermarket simply by slotting the card into the reader; no subsequent PIN was required. Whilst this problem was quickly identified as a possible issue and rectified, many other examples exist, for example authentication mechanisms that are disabled by default. Mobile phones and until more recently wireless access points were both shipped without any authentication enabled. Given that both are consumer goods with widespread appeal, many users would not even have been aware that authentication was a good idea, let alone know how to enable it.

To further complicate the problem, many scenarios exist where point-of-entry authentication unfortunately falls short of providing sufficient protection. Systems, technologies and applications have been developed that enable users to access a far richer environment of information, yet little consideration has been placed on how to best protect it. For instance, access to a mobile phone is initially protected through a PIN code (typically); however, the phone is then able to remain on and require no further revalidation of the user. Moreover, during the session, the user is able to do everything from playing a game to accessing all the personal information stored on internal memory, without any consideration to the privacy and security of the data.

System-based authentication is very common and is what we experience when accessing computers generally. The authentication approach has been deployed to determine permission over access to the device or operating system (OS) itself and provides no protection for accessing the services and information stored on the device (although some services will require further authentication credentials such as banking applications on the Apple iPhone). But access to the device itself does not actually provide the user with any actual data or misuse. Indeed, it is only when the user subsequently accesses an application or a document that misuse potentially takes place. An argument exists therefore that current authentication in many cases is not actually linked to the accesses that require protection. If a user were to use a computer merely to find the time, play a game of minesweeper or to use the calculator, is it necessary to authenticate them?

The desire for a closer tie between the authentication mechanism and the actual data to be protected is useful because:

1. Authentication mechanisms are intrusive and inconvenient to users – forcing users to authenticate when perhaps there is no actual need seems counterproductive.
2. A better understanding of what needs protecting will enable a better deployment of authentication approaches. Why is an identical PIN code able to confirm a transaction of £1 and £10,000? The risks and subsequently the impacts associated with both transactions are completely different.

Reflecting upon the security countermeasures in place already, it is clear that providing access control mechanisms to secure access to the service or application seeks to ensure that only the validated user is given access. However, the assumption of the authorised user has frequently been determined some time previously. The problem that arises is that effective systems security then relies upon accountability countermeasures to identify and recover from subsequent misuse. But this happens after the fact – merely protecting systems from that attack happening again. If authentication could successfully be achieved just prior to access control, it would provide a means of providing proactive security, rather than the reactive security that accountability enables.

3.3 Risk Assessment and Commensurate Security

To what degree does user authentication actually meet the desired level of security? In the examples given in the previous section, it appears little thought had gone into what level of security was required, but more focus was placed on ease of implementation and cost-effectiveness. Ensuring commensurate security, where the countermeasure in place is commensurate with the risk to the asset it is protecting, appears a logical step in providing more effective security, where effective is a measure of both the security provided and the user convenience. Commensurate security is nothing new, and information security standards, such as ISO27001 (2005), have been advocating it. The process of Risk Assessment enables organisations to ensure they mitigate the information risks to an acceptable level through the deployment of countermeasures to protect assets.

The process of risk assessment is defined as 'the assessment of threats to, impacts on and vulnerabilities of information and information processing facilities and the likelihood of their occurrence' (ISO27001, 2005), and involves the identification of assets that need to be protected and the identification of threats and vulnerabilities related to those assets. An illustration of the process is shown in Fig. 3.1.

It is also important to ensure risk assessment processes are repeated periodically in order to ensure the most appropriate technology is still deployed. The problem that arises here is frequently due to function creep and evolving organisational priorities;

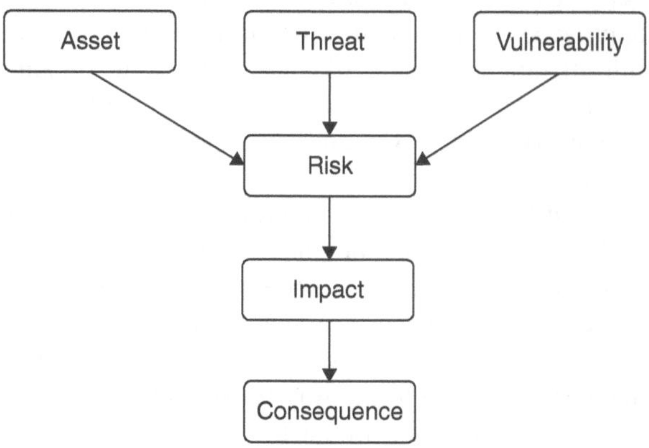

Fig. 3.1 Risk assessment process

current countermeasures fail to meet current levels of risk, but the costs of putting these right are outweighed by the ability to mitigate the risk through other means – such as purchasing insurance. Subsequently, we have to live with systems that fail to provide effective protection. The old magnetic strip–based credit cards are an interesting illustration of this. Card cloning through skimming was easily achievable for many years before the technology was replaced; yet financial institutions took a considerable amount of time to update the technology. The decision was simply cost-effectiveness – up until the technology was superseded, it was cheaper to deal with the level of fraud than to role out a new technology.

Risk assessment has been identified as a useful approach to achieving commensurate security. However, it is worth highlighting that this is only useful to organisations that can afford to undertake such assessments – they tend to be expensive, time-consuming and require experienced auditors. For members of the public or small organisations, these approaches are out of their reach, so ensuring that appropriate authentication approaches are utilised is a far more challenging task. Consider, for example, an individual who uses a home computer. Will that individual undertake an assessment of their system to appreciate the risks involved and ensure appropriate countermeasures are in place to mitigate those risks? In the majority of scenarios, they would not. Whilst system designers can certainly have a role to play in the applications they design – ensuring appropriate technologies are deployed – they do not always know for what purpose the application will be used (e.g. using excel to view banking transactions) and have little control over the end-user. Significant consideration of the user is therefore required to ensure they are able to appreciate the risks involved with technology.

Risk assessment is therefore a time-consuming and costly process but does provide a useful basis for understanding the information security risks your assets pose. From an authentication perspective, risk assessment completed correctly can identify the most appropriate technology solution for a given environment. The problem that arises is that too frequently risk assessment is not performed particularly well, or assumptions (such as passwords are effective) are relied upon which turn out to be false. Moreover, many solutions to the authentication problem are solved through a blanket-based approach, with a single-authentication solution deployed throughout the organisation or technology. For instance, single sign-on approaches seek to reduce the burden upon the user. However, it also assumes all information accessed is of equal value to the organisation, from accessing a computer, to copying a database of customer details. Little consideration is given to the varying risks that exist and ensuring appropriate authentication mechanisms are put in place.

To further illustrate the point, consider the protection required by accessing several services on a mobile phone. It is substantially different to that required by a bank account. Figure 3.2 shows a 3-dimensional representation of how current authentication schemes deal with security, keeping a single level of security for all services. Figure 3.3 shows how the threat derived from each service could add another dimension to the way in which the security level is defined. Each service carries a certain risk of misuse, and this ought to be a factor in deciding the appropriate level of security.

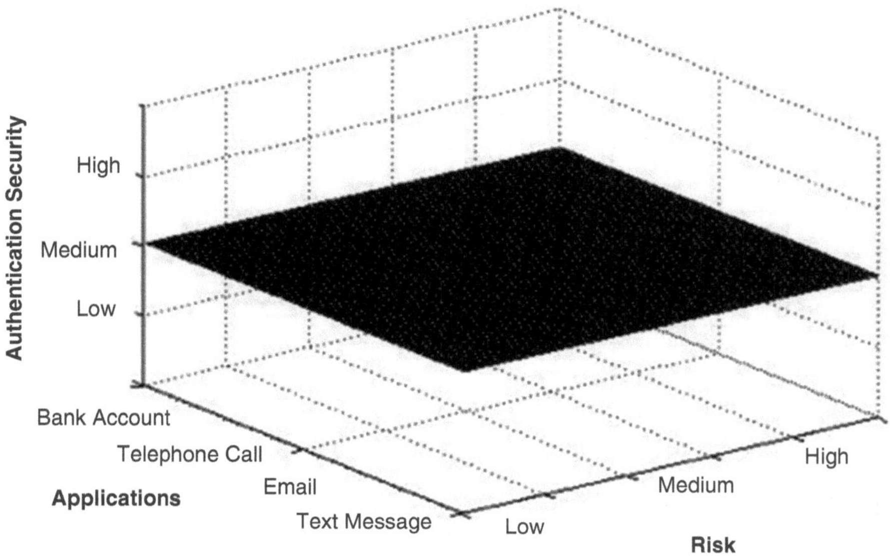

Fig. 3.2 Authentication security: traditional static model

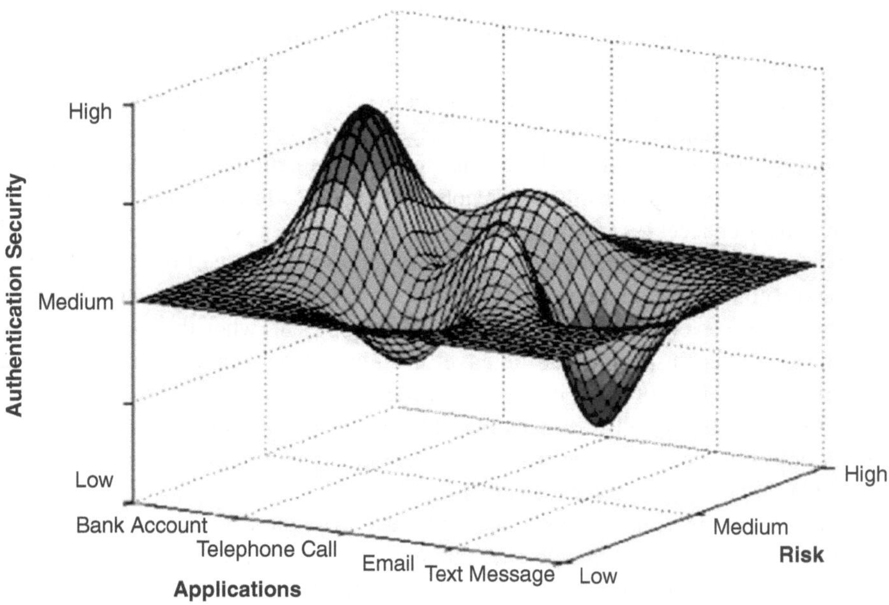

Fig. 3.3 Authentication security: risk-based model

The level of security is more appropriately assigned to each service, so that each service or function can independently require a certain level of authentication (and consequently trust in the legitimacy of the user) in order to grant access to the specific service. In this way, more critical operations can be assigned greater protection, leaving less risky operations to a lower level of trust.

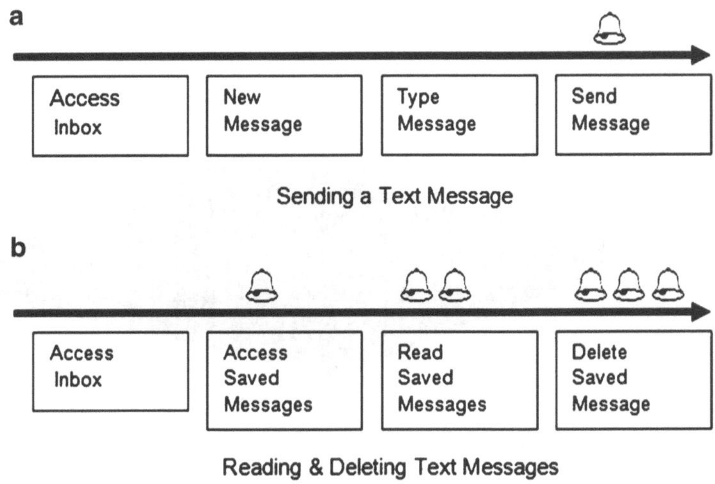

Fig. 3.4 Variation of the security requirements during utilisation of a service. (**a**) Sending a text message, (**b**) Reading and deleting text messages

This logic can also be applied a stage further. It can also be argued that the level of security within a service or application is likely to change during the process, as key stages will have a greater risk associated to them than others. In order to carry out a specific task, a number of discrete steps are involved, which may not carry the same level of sensitivity (i.e. some processes are more critical, whereas others are simply operational steps that assist in the completion of the desired task). A simple example that illustrates this notion is the procedure of accessing an email inbox. The user accesses the inbox and at that instance there is no real threat involved as the operation cannot lead to any misuse in its own right (see Fig. 3.4a). Even if the next step is to create a new message and start typing the content, no additional risk exists. However, the security implications actually start when the user is pressing 'Send' as it is at this point that the adverse impacts from impostor actions would actually begin. By contrast, in Fig. 3.4b, the user again accesses the inbox, but tries to access the saved messages instead. This time the requirement for greater protection occurs earlier in the process, as accessing the saved messages could affect confidentiality if an impostor reads them.

Thus, it can be seen that each operation has different sensitivities and as such each step of the process changes the threat and therefore the risk level. Enabling both *inter* and *intra* process security would permit a far more robust approach to ensuring commensurate authenticity of the user. In order to apply individual security levels to applications and services there is a need for threat assessment to classify the security risks associated with them, from both organisational and individual perspectives. From this classification, a security level could be attributed to each type of service, and subsequently to the level of trust required in the legitimacy of the user.

3.4 Transparent and Continuous Authentication

Point-of-entry authentication is a proven and useful approach in scenarios where one-off intrusive verification of the user is necessary, for example, passports and border control. In scenarios where successful access pertains to a continued set of access privileges, point-of-entry authentication merely provides initial confirmation of identity but not ongoing persistent verification, for instance, use of a mobile phone. Typically protected through a PIN code initially, once entered the device can remain on indefinitely without any re-authentication of the user. Moreover, that initial authentication does not actually pertain to any risky operation; it merely provides subsequent access to what does need protecting. In such scenarios, re-authentication of the user would be useful to ensure continued validity of the user.

Continuous re-authentication does, however, introduce some sort of a contradiction between the need to more closely tie authentication with specific access and overloading the user with the need to authenticate. Requiring the user to authenticate in order to access a sensitive service is likely to increase the number of occasions the user needs to authenticate and thus increase the burden upon and inconvenience to the user. Earlier chapters have already identified the significant authentication burden placed upon users and any new approach to solving the problem should seek to reduce not increase this burden.

Furthermore, given that none of the authentication approaches provides 100% accuracy, how reliable is one-off authentication? The concept of a Boolean decision is simply not a true reflection of the situation. It takes no account of the likelihood of error (which in some circumstances is very high). A more realistic measure is therefore required which takes into account not only whether the authentication result was true or false but also what authentication technique was used, how secure it is and how confident it was in the decision it made. For instance, currently from an output perspective, the decision from a password-based authentication takes no consideration of the strength of the input. The password for one user might be very weak, another very strong, but the system would still permit the same level of access to each user.

From an analysis of the authentication requirements, it is evident that the following objectives exist:

1. Reduce the authentication burden upon the user
2. Improve the level of security being provided
3. More closely link authentication of the user with the subsequent request for access
4. Take into account that a commensurate authentication approach is utilised depending upon the risk associated with the access request
5. Provide a more effective measure of identity confidence that goes beyond a simple Boolean decision

This is no simple set of requirements to meet. To improve the level of security but at the same time reduce the burden upon the user are at first glance almost

Fig. 3.5 Transparent authentication on a mobile device

contradictory in nature. The requirements also highlight a significantly different perspective on authentication. Rather than determining authenticity with a (falsely) perceived certainty, the proposed requirements suggest that the system is never completely certain whether the user is indeed the authorised user, but simply that it is $x\%$ confident that it is.

The key to enabling re-authentication, reducing the burden upon the user and increasing security is the ability to move away from intrusive authentication to non-intrusive or transparent authentication. If authentication can be performed in the background, without a user having to explicitly provide a sample, but rather capturing the authentication sample during the user's normal interaction with the device, the user would not experience any inconvenience – as they would be unaware that authentication is taking place. If the authentication is happening transparently, it is possible to perform authentication on a continuous basis and thereby enable both re-authentication of the user and provide a closer tie between authentication and specific access requests. Furthermore, the multiple authentication decisions that result will provide stronger confidence in the authenticity of the user and therefore improve the level of security.

Theoretically, transparent authentication can be achieved by any authentication approach that is able to obtain the sample required for verification non-intrusively. Biometric techniques fall within this category, such as capturing photographs of the user whilst teleconferencing and subsequently using facial recognition, or perhaps recording their voice whilst they are in telephone conversation and using speaker recognition. Certain token-based approaches could also be used in this context. For example, contactless-based tokens in particular require little interaction from the user, merely close proximity to the reader to authenticate. Secret knowledge–based approaches are, however, unsuitable, as they can only be utilised in an intrusive fashion.

In order to appreciate how this might work in practice, Fig. 3.5 illustrates the capabilities of a common mobile device. In particular, it highlights the inbuilt hardware that exists that would enable authentication samples to be captured. When the

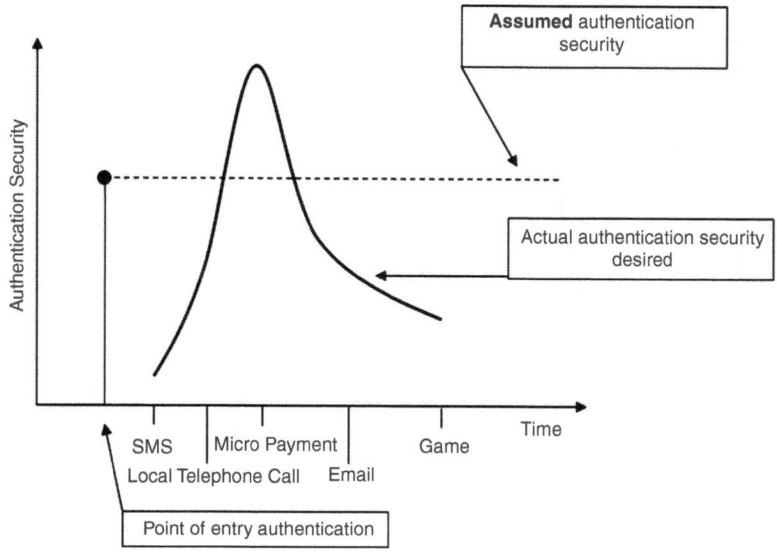

Fig. 3.6 Normal authentication confidence

user interacts with a technology, it is possible that some form of biometric-based sample can be captured. For instance, whilst typing, built-in front-facing mobile phone cameras have the ability to capture the person looking at the screen. Keystroke analysis, a biometric approach based upon typing characteristics, could also be utilised. Whilst the user is surfing a web page, service utilisation – a user-profiling approach – could be used. Whilst making a telephone call, voice verification could be applied. A similar set of hardware features exist on other computing platforms such as laptops, desktop PCs, PDAs and tablet PCs.

Whilst sample capture is merely one stage of an authentication process, with other issues such as processing, storage and computational complexity all playing a role – especially with the more processing limited mobile devices – it is the key phase to establishing transparent authentication. Depending upon the authentication architecture, many of the other aspects can always be outsourced to other systems (Part 3 describes possible architectural considerations and operational decisions).

The move from one-off authentication decisions to a transparent and continuous mechanism provides the system with a continuous rather than discrete level of confidence in the authenticity of the user. Figures 3.6 and 3.7 illustrate the difference in approaches, with the former demonstrating infrequent authentication with subsequently little validation of the authenticity of the user and the latter a time-dependent, authentication-dependent measure of confidence of authenticity. Whilst both graphs are merely indicative of what might happen rather than real-life traces, it is clear that the continuous confidence measure would enable a higher level of security. With each interaction, the system is able to obtain a better confidence in the identity of the user. More importantly, however, it provides a realistic measure of confidence rather than providing a false sense of security.

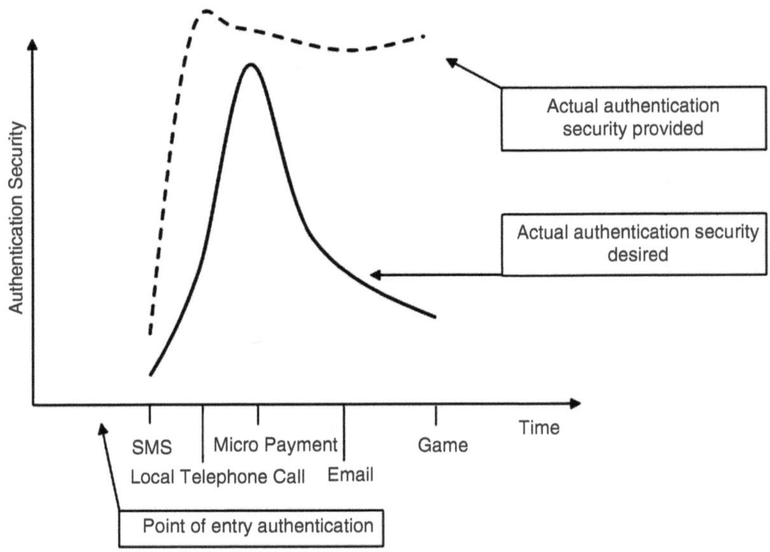

Fig. 3.7 Continuous authentication confidence

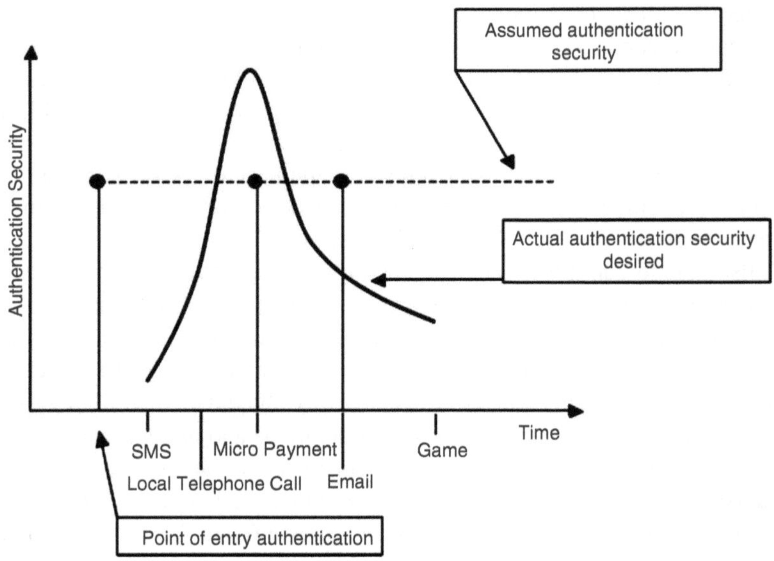

Fig. 3.8 Normal authentication with intermitted application-level authentication

Even if we consider the trace depicted in Fig. 3.8, it is clear that whist additional point-of-entry intrusive authentications for specific applications or services will provide a good level of confidence for the specific application or service it is related to, it fails to enable other services and applications to take advantage of that confidence. In addition, it still remains intrusive and therefore fails to reduce the inconvenience and burden upon the user.

Transparent authentication is not a solution to all authentication problems. It can resolve issues of inappropriate selection of authentication techniques or poor implementation and policy decisions. It is also only appropriate for particular environments – specifically computing related – where persistent access is required. It is also worth pointing out that transparent authentication is not about covert security. It is not about authenticating the user without their knowledge or specific consent. The concept is about overt security, making individuals aware of what the system is doing and how the user will benefit from the increased protection it provides.

Transparent authentication also introduces its own set of complications and issues. Increasing the number of authentication techniques available on a device and the number of authentications that occur will increase the cost and complexity of the system. However, whilst today's intrusive point-of-entry authentication mechanisms might appear simple and cheap in comparison, it is worth considering to what degree do those costs truly reflect all the indirect costs when the underlying assumptions that ensure security fail: the cost to an organisation for a breach of confidential information; the cost to an individual for the misuse of his or her mobile phone and the cost to financial organisations for the misuse of online banking services. In many of these scenarios, the true cost of deploying a security countermeasure and the resulting benefit it brings in preventing a security incident is not considered.

3.5 Summary

What is currently being achieved by authentication approaches to date varies from carefully considered implementations to inappropriate deployments that are overly concerned with the financial implications rather than the security concerns. In the forthcoming years, the continuing misuse of systems and increasing level of cybercrime, legislation and regulation and a demand from individuals themselves will force system designers to reconsider their approaches. Designers will have to consider security, usability, applicability, universality and costs (both of implementation and of breach) and deploy more appropriate technologies. Simply evolving current authentication mechanisms to fit new problems without a considered re-evaluation of the system will not be sufficient.

Within this remit, transparent authentication has a significant role to play in enabling strong, convenient and continuous authentication of the user. Authentication is no longer a one-off process but an ongoing confidence in identity assurance that truly reflects the security being provided. Underlying assumptions such as policies being adhered to, loss of tokens being reported and sufficient complex passwords being utilised are all unnecessary. Indeed, the reliance upon the human aspects to ensure security is almost effectively removed, providing a more realistic and reliable measure of identity verification.

Despite advocating the use of transparent authentication to solve the authentication issues that currently exist, there are a number of new problems that will need to be considered. The remainder of this book is focused upon identifying and analysing these issues. Part 2 focuses upon the core authentication techniques themselves and their applicability to being applied in a transparent fashion. Part 3 examines the architectural considerations with deploying such an approach.

Reference

ISO: ISO/IEC 27001:2005 Information technology – Security techniques – Information security management systems – Requirements. International Standards Organisation. Available at: http://www.iso.org/iso/catalogue_detail?csnumber=42103 (2005). Accessed 10 Apr 2011

Part II
Authentication Approaches

Chapter 4
Intrusive Authentication Approaches

4.1 Introduction

Authentication has been at the cornerstone of information security since the inception of information technology (IT). Whilst the foundations upon which they rely have changed little, technology has evolved and adapted these approaches to fit a variety of solutions. Prior to describing the nature of transparent authentication, the current technological barriers to implementation and the advantages such an approach could have, it is important to establish a baseline understanding of the current nature of authentication, the current technological requirements, limitations and deployments. From such a basis it is possible to better appreciate the unique environment within which transparent authentication operates and the benefits it could bring.

A technical overview of the authentication approaches is also required to understand the current state of the art and appreciate what advances have been made. It will enable a range of technical nomenclature related to authentication to be introduced that will allow for comparisons in the operation and performance of authentication techniques to be drawn.

4.2 Secret-Knowledge Authentication

In essence, all authentication approaches result in some form of secret knowledge–based authentication. A token is merely an electronic store for a password string users would find difficult to remember. A biometric token simply contains a unique biometric signature of the user, which is again, in electronic form, simply a unique password associated to that user. The inability of people to remember sufficiently complex passwords that would enable security has been identified for almost as long as the initial conception of using a password. Numerous approaches and studies

N. Clarke, *Transparent User Authentication: Biometrics, RFID and Behavioural Profiling*, DOI 10.1007/978-0-85729-805-8_4, © Springer-Verlag London Limited 2011

have been conducted examining various approaches and mechanisms that allow designers to capitalise upon the relevant ease of implementation and good levels of end-user knowledge and experience on how they work.

The following sections will examine the current approaches to secret-knowledge authentication, including an investigation of graphical passwords that seek to take advantage of our human ability to better recognise something rather than recalling it.

4.2.1 Passwords, PINs and Cognitive Knowledge

The basis for any effective authentication approach is to have a unique code that an attacker would not be able to obtain – through prediction, guesswork or brute-forcing (trying every possible combination).[1] Whilst brute-force attacks exist for all three authentication categories – secret-knowledge, tokens and biometrics – prediction and guesswork are solely secret-knowledge attack vectors. They focus on human nature and our inability to select appropriate passwords. Common compromise includes:

- Use of dictionary words – including very predictable passwords such as 'love', 'sex', 'password', 'qwerty', '123456' (recall that Table 1.4 highlights the 20 most common passwords utilised from a couple of studies).
- Passwords derived from personal information, such as a date of birth, spouse's name, close family, where you live.
- Use of simple deviations of words and/or numbers.

The use of word deviations is at least an attempt by a user to try to obfuscate their password. For example, 'luke23', could be the combination of a name with an age. Whilst users perceive this would be secure, freely available attack tools have functionality built in to try such combinations before attempting the more arduous task of a brute-force attack.

Furthermore, users' poor use of passwords only exasperates the issue, with users ignoring policy and good practice guidelines. Examples of compromise include:

- Passwords that are too short and therefore reduce the number of attempts required.
- Writing them down (e.g. on notes which are then placed under the keyboard or stuck on a monitor).
- Sharing them with colleagues.
- Not renewing the password on a frequent enough basis.
- Using them on multiple systems so that a compromise on one system would result in a compromise across many systems.

[1] Whilst other attack vectors exist, the emphasis at this point is not on attacking systems or protocols to recover the password. This will be examined in Sect. 4.2.3.

Table 4.1 Password space based upon length

Length	Password space Numeric	Alphabetic	Alphanumeric	Alphanumeric + Symbols
1	10	26	36	46
2	100	676	1 K	2 K
3	1,000	18 K	47 K	97 K
4	10 K	457 K	2 M	4 M
5	100 K	12 M	60 M	206 M
6	1 M	308 M	2 G	9 G
7	10 M	8 G	78 G	436 G
8	100 M	208 G	3 T	20 T
9	1 G	5 T	102 T	922 T
10	10 G	141 T	4 P	42 P
11	100 G	4 P	132 P	2 E
12	1 T	95 P	5 E	90 E
13	10 T	2 E	171 E	4 Z
14	100 T	64.5 E	6 Z	190 Z

Key: M = Mega = 10^6, G = Giga = 10^9, T = Tera = 10^{12}, P = Peta = 10^{15}, E = Exa = 10^{18}, Z = Zeta = 10^{21}

Inherently, if users did not fall foul of the aforementioned issues, secret knowledge–based approaches can indeed be very effective, with their level of protection directly linked to their length. For example, Table 4.1 illustrates the effective password space – the number of possible permutations of a password – for varying lengths of password. With a ten-character password comprising numbers, letters and symbols, an attacker would have over 42,420,747,482,776,600 combinations to try. This is a fairly insurmountable attack even with password-cracking tools. Table 4.1 also illustrates why short passwords are ineffective at providing any real level of security.

The number of permutations for a password is also referred to as *Entropy* (Smith 2002). Whilst Table 4.1 illustrates the number of possible permutations of passwords in real numbers, as the password length increases the resulting number is very unwieldy to use in practice. Therefore, it is more common to measure entropy in terms of bits. This practice is well established in the cryptographic domain, where terminology such as 128-bit or 256-bit keys are used to illustrate the strength of the key being used. The number of bits represents the amount of random-access memory (RAM) required to store the number and therefore the resulting number is always rounded up to the nearest bit. The practice of using bits also makes it easier from a comparison perspective. The formula used to compute the bit space is:

$$\text{Entropy}\left(\text{bits}\right) = \log_2(\#\,\text{of permutations})$$

For instance, the bit space of the 10-character alphanumeric password including symbols is 56 bits. Table 4.2 illustrates the entropy of passwords with varying password length and composition.

Table 4.2 Password space defined in bits

	Password space in bits			
Length	Numeric	Alphabetic	Alphanumeric	Alphanumeric + Symbols
1	4	5	6	6
2	7	10	11	12
3	10	15	16	17
4	14	19	21	23
5	17	24	26	28
6	20	29	32	34
7	24	33	37	39
8	27	38	42	45
9	30	43	47	50
10	34	48	52	56
11	37	52	57	61
12	40	57	63	67
13	44	62	68	72
14	47	66	73	78

Table 4.3 Examples of cognitive questions

Cognitive question
What is your mother's maiden name?
What is your favourite movie?
What is your favourite colour?
What city were you born in?
Who is your favourite music artist?
What was your first pet's name?

The bit spaces reported in Table 4.2 give a measure of the entropy assuming all possible combinations have equal weighting. In many passwords, a bias exists towards the use of certain words or combinations and this would result in a smaller bit space than reported here, for instance, the use of a dictionary word rather than a random set of characters. The English Oxford Dictionary has 171,476 words (Oxford University Press 2010), which relates to 18 bits of entropy. A random 4-character password (including numbers and symbols) has 23 bits of entropy.

Personal identification numbers (PINs) also fall into the same category as passwords from an entropy perspective, but are arguably worse off due to the smaller set of possible combinations. For a typical 4-digit PIN, the entropy is only 14 bits (or 10,000 permutations). However, cognitive-based approaches fair the worst with frequently very limited bit space.

Cognitive-based password approaches seek to remove the problem of having to remember a password that has no direct association to you but instead relies upon asking you a number of questions that pertain to your personal history, likes/dislikes, opinions and attitudes. Table 4.3 illustrates a series of examples of cognitive questions.

Table 4.4 Typical password policies

Policy	Description
Minimum length	Length must consist of at least a predefined minimum number of characters. Typically this is now greater than 9
Composition	Password must consist of a combination of characters, numbers and symbols
Case-sensitive	Use of at least one upper-case character
Aging	Change your password on a frequent basis. This can vary considerably depending upon the environment. A monthly change is not uncommon
Reuse	When resetting passwords, previously used passwords should not be used
Multiple systems	Passwords should be unique to each account you have. Identical passwords on multiple systems should be avoided
Handling	Passwords should be handled with care. They should not be written down or shared with colleagues

The nature of the questions tends to be either factual or opinion-based. Whilst the former is useful from a longevity perspective – facts are unlikely to change over time, whereas opinion is – facts are also simpler to establish in many cases. For instance, determining your mother's maiden name only requires a quick search on an ancestry web site. Where you were born and what was your first school, amongst many others questions, could be established through social networking sites such as Facebook.

Given the responses are dictionary-based, it is not sufficient to merely use one question and frequently they require the user to successfully answer a series of cognitive questions (or potentially use a cognitive-based question in addition to another measure). So whilst reducing the need for the user to remember the password, the approach does require them to enter several responses, increasing the time to authenticate (and potentially the burden). The largest barrier to cognitive-based approaches is the entropy. Each question will only have a finite and fairly small set of possible answers. Whilst this issue is compounded, by requiring the user to answer multiple questions simultaneously (thereby removing the opportunity to attack each in turn), the availability of personal information online, social engineering techniques and the predictability of responses tends to provide a relatively low level of security. For instance, what is your favourite food? If hypothetically there were 20,000 words for food (representing 11% of the total number of words in the English dictionary), this would only represent 15 bits of entropy. Moreover, of those 20,000 words, a far smaller subset is likely to be more regularly used than others (e.g. pizza, burgers, chips, fries and lasagne) and therefore the entropy will always be far smaller.

Given the deficiencies in password-based approaches, various additional measures have been taken to try and ensure (rather force in some cases) users to utilise stronger passwords. Organisations have put policies in place to ensure users meet the minimal level of protection and most modern OSs support some form of policy enforcement. Table 4.4 illustrates some of the common password policies applied.

Fig. 4.1 Googlemail password indicator

Various service providers have also enforced particular password policies, such as password composition and length. However, service providers have to be careful they don't overburden the user as this will result in a loss of custom or a significant increase in customer service requests to reset the password. A more recent evolution to educate and enable the user to select more effective passwords is the password strength indicator. Figure 4.1 illustrates Google's version of the indicator. Whilst an extremely novel and potentially useful feature, the lack of a standard approach between vendors in defining what is weak versus what is strong means each has its own (and subsequently different) policy to define what is a weak and strong password, which gives users mixed messages about the strength of a password. For example, the strong password strength in Fig. 4.1 was obtained by merely typing 'nathanclarke' as the password – clearly a weak password. Googlemail is by no means alone, with few password indicators providing an effective measure of security. Interestingly, this one password permits access to a range of Google's services, from email, calendar and documents – further illustrating the consequence of having the identical security countermeasure protect numerous assets.

Online banks are an interesting example for discussion as they represent one of the few services online that end-users can clearly identify as needing security and as such can afford to deploy more effective security. In general, the approaches they utilise are:

- An ID number. Whilst not secret like a password, advice from the bank requests you keep it private.
- A personal piece of information, typically a date of birth, but other cognitive based questions are also asked, such as favourite film, favourite place, place of birth and favourite food.
- A form of secret-knowledge password. In some cases, such as HSBC, this is a 6–10-digit number; for others such as Barclaycard, it is a variable-length alphanumeric string.

They also have different policies on resetting forgotten passwords. HSBC requires you to call its Internet helpdesk. Barclaycard requires you to enter details contained on the credit card (such as the card number, expiry date, etc.) in addition to two letters from your memorable word. So they utilise a range of secret, non-secret and cognitive-based approaches, rather than depending upon one.

Interestingly, however, banks have also begun to use the cash card as a means of verifying particular actions on the online account. NatWest has distributed card readers to online customers who need to use the reader with the card to perform certain functions such as setting up a new payee on the account. Money can be set to previous payees but new payees must be set up more securely the first time of use. This clearly demonstrates a two-level access approach to online banking and reflects the lack of trust in the aforementioned approaches. One permits a level of access to effectively read information and a limited degree of modification. However, for more substantial operations such as transactions of money from the account to unknown accounts, additional one-time passwords that need both the card reader and debit card are required.

Whilst the one-time password accompanied with a secret-knowledge approach is a secure operation, it is also arguably a fairly expensive and potentially restrictive one. Everyone who has an account would need to be supplied with the necessary card reader and they would also have to remember to take it with them if they wish to authorise transactions from another location (e.g. from work rather than home). Furthermore, whilst this might be appropriate for one bank, what happens when you have multiple accounts across a range of providers – individual card readers from each bank?

4.2.2 Graphical Passwords

The inability for people to remember a string of characters of sufficient length and complexity has given rise to graphical passwords, an approach that relies upon the ability to more reliably recognise and recall images over characters. Studies have confirmed that people generally have a far better ability to recall and recognise pictorial representations (Paivio et al. 1968; Shepard 1967). This has led to a number of research studies and commercial offerings that rely upon the use of some form of graphical representation. The work can be broadly classified into two categories:

- Recognition – identifying an image or series of images that have been previously seen, for example, a choice-based approach depicted in Fig. 4.2. The password is comprised of a series of images that must be selected in order, with each click providing a new set of images to select from. The arrangement and placement of images is randomised to minimise shoulder surfing and other attack vectors.
- Recall – recreating an image that has been previously drawn. This is similar to the process of password authentication but focused upon imagery rather than

Fig. 4.2 Choice-based graphical authentication

textual passwords, for example, a click-based approach that requires the user to recall and select a series of points on an image (as illustrated in Fig. 4.3).

The key advantages of graphical-based authentication are an increase in the usability and convenience from an end-user perspective. They also have theoretically fairly high entropy when compared to passwords (that are actually used in practice). Blonder (1995) identified in his patent that his click-based approach could have 13.6 million combinations (equivalent to an entropy of 24 bits).

A number of commercial products and implementations exist, most notably Passfaces (2011), as illustrated in Fig. 4.4. Rather than using generic images, this choice-based approach utilises human faces – which relies upon the premise that humans are better at recognising faces than generic objects. This particular approach requires the user to select the appropriate face three times from a new set of nine images.

Compared to password-based approaches, the time taken to enrol the user and subsequently to authenticate them is increased. However, studies have demonstrated that graphical passwords can be more effective than their textual counterparts. A study by Brostoff and Sasse (2000) highlighted that users using the Passfaces approach were a third less likely to experience a login failure than text-based users, even though users had a third of the frequency of use (the less frequently an approach is used, the more likely the user will forget the password).

Fig. 4.3 Click-based graphical authentication

Fig. 4.4 Passfaces authentication

Graphical techniques to authentication have also found their way into online banking applications. These implementations are primarily motivated from the threats that Phishing poses and they present an image to the user when logging into their online account which signifies the legitimate bank over a Phishing web site. The bank is able to display the image if it recognises the computer from which the user is logging in. Should the user log in from a different location, the system will take additional security measures such as asking additional questions. Whilst a Phishing site could of course present an image, it would not be able to display the correct image that the user had initially selected. SiteKey used by the Bank of America is a typical example (Bank of America 2011).

4.2.3 Attacks Against Passwords

Attacks on secret-knowledge approaches can be broadly classified into technical and non-technical approaches. Technical approaches include:

- Brute-force attack – a systematic approach of trying every permutation of password.
- Phishing – an email-based threat that purports to have come from a legitimate source (e.g. a bank) asking you to reconfirm your authentication credentials. Whilst the attack itself is not technical, it requires a certain degree of technical expertise to set up the supporting system to capture the information.
- Manipulated input device – capturing the information as close to the source as possible through fake input devices or devices that intercept such information. Hardware-based keyloggers operate as an inline device between the keyboard and the input connection on the computer.
- Trojan Horse – a malicious software-based approach to the capture of login information.
- Network sniffer – promiscuously capturing all traffic on a network and analysing it for password-based information, particularly from cleartext protocols, but hashed passwords are also useful for subsequent cracking.
- Electromagnetic emissions – studies have demonstrated the ability to determine key presses by listening to the electromagnetic signals emitted when pressing keys from distances of up to 20 m (Vuagnoux and Pasini 2009).

Non-technical attack vectors include:

- Guesswork – whilst seemingly unlikely, depending upon the system, the knowledge the attacker is able to obtain and the strength of the password utilised, it may be possible to obtain the password through informed guessing.
- Shoulder surfing – observing the password being entered from a close proximity.
- Social engineering – manipulating the target user into releasing the necessary login information. It can also include the lesser threat of revealing information about themselves that would assist in a guesswork or as a seed to a brute-force attack.

Whilst technical attacks might have initially been the preserve of those with appropriate knowledge or skill, the widespread use of the Internet has made technical attack vectors possible for anyone to use. Hardware keyloggers are available from providers on the Internet from £25 to £40. The ability to perform network

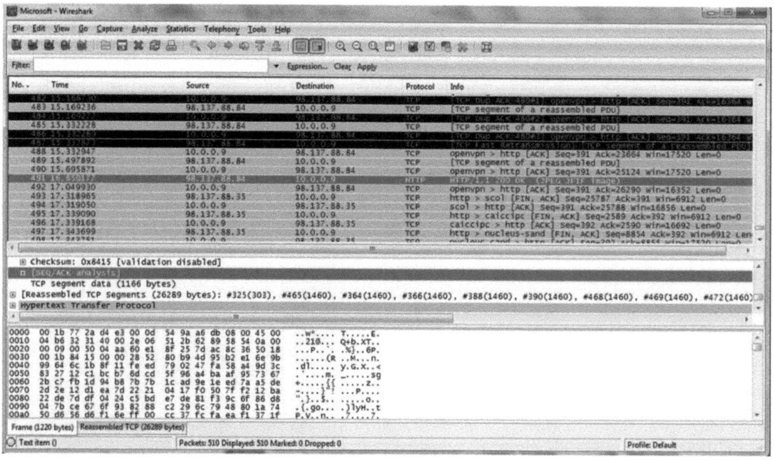

Fig. 4.5 Network monitoring using Wireshark

Fig. 4.6 Senna Spy Trojan generator

sniffing is simple using freely available network-monitoring software such as Wireshark (Wireshark 2011). Figure 4.5 illustrates Wireshark sniffing network traffic. Even Trojan Horses can be created without knowing how to write a single line of code (Symantec 2007). The Senna Spy Generator, as illustrated in Fig. 4.6, is a

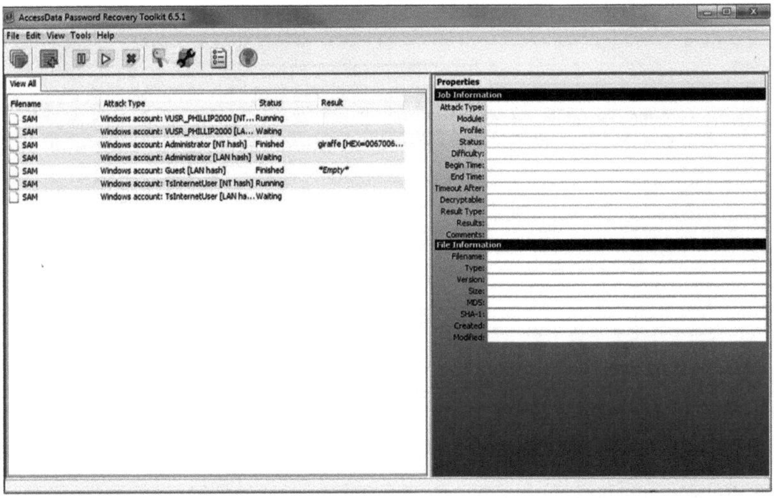

Fig. 4.7 AccessData password recovery toolkit

program that will enable a hacker to simply create variants of the Senna Spy Trojan Horse by following the simple on-screen instructions.

The threat that has received most focus, however, is the password cracker. Whilst ultimately converging to a brute-force attack, various tools have varying levels of sophistication that assist it in identifying the password. The ability to crack a password depends upon the form it is stored/communicated in, which in itself relies upon the cryptographic algorithm utilised. In many cases, the demand for these tools has also been driven from a legitimate need rather than anything criminal. System administrators have a legitimate desire to ensure their users are using strong passwords. There is also a legitimate need by law enforcement agencies to be able to recover information that is encrypted. Three frequently cited tools are:

- AccessData's Password Recovery Toolkit (PRTK). A tool created from the computer forensics domain and the need to decrypt files that could be pertinent to an investigation. The tool includes the functionality to recover passwords from over 80 applications, including the Windows SAM file (the file that contains a user's login password for Windows). In addition to having brute-force and multilingual capabilities the tool is also able to generate its own dictionaries based upon every permutation of strings stored on the investigating system. This approach is capitalising on the assumption that the password is based upon something that might appear somewhere else within your system. An illustration of PRTK is shown in Fig. 4.7. Whilst personalised dictionary attacks take relatively short periods of time to complete, brute-force attacks of passwords can take a considerable amount of time.
- Ophcrack. A Windows password cracker designed to crack LM and NTLM hashes. These hashes store the Windows password information in the SAM file.

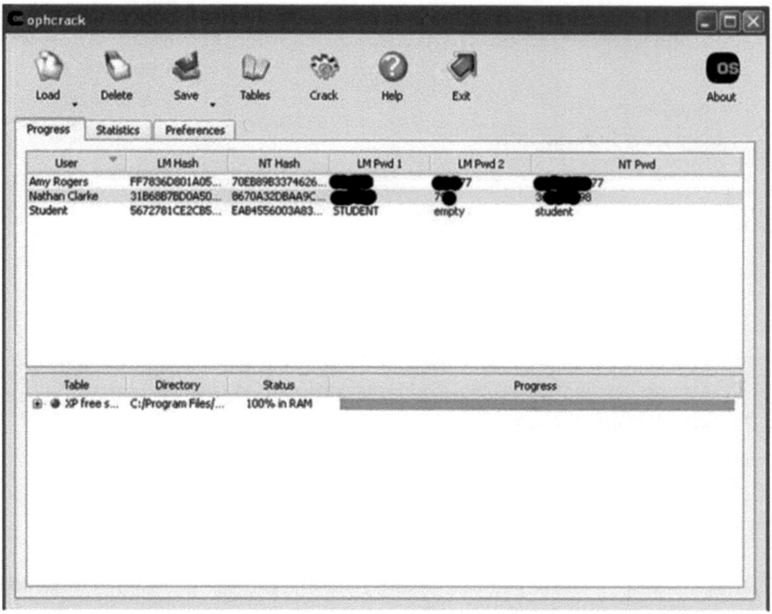

Fig. 4.8 Ophcrack password recovery

In comparison to PRTK, however, Ophcrack utilises Rainbow tables, which significantly reduces the time taken to recover the password (in most cases) through the use of pre-computed tables. Whilst a discussion of Rainbow tables is outside the scope of this text, please refer to Oechslin's paper on 'Making a Faster Cyptoanalytic Time-Memory Trade-off' (2003). As illustrated in Fig. 4.8, Ophcrack is able to recover ten-character passwords within minutes.

- Cain and Abel. A Microsoft Windows–based password recovery tool capable of cracking a variety of passwords from sniffing network protocols as well as LM and NTLM hashes. Possibly the most comprehensive password recovery tool freely available, Cain and Abel, is able to use a variety of mechanisms to recover passwords including dictionary, Rainbow tables and brute-forcing. An illustration of the functionality is presented in Fig. 4.9.

The traditional textual-based password is not the only approach open to compromise. Graphical passwords also suffer from a number of weaknesses. The choice of image(s) utilised in graphical passwords play a significant role in how strong the approach is. For click-based approaches, poorly selected images have particular areas that people are more likely to select than others. These common areas subsequently reduce the entropy of the approach as the password is focused upon a series of common areas. A study by Thorpe and van Oorschot (2007) reported that in a 31-bit space 36% of passwords could be recovered (12% within 16-bit entropy). Even when the image is chosen well, 'hotspots' of activity can appear within the image.

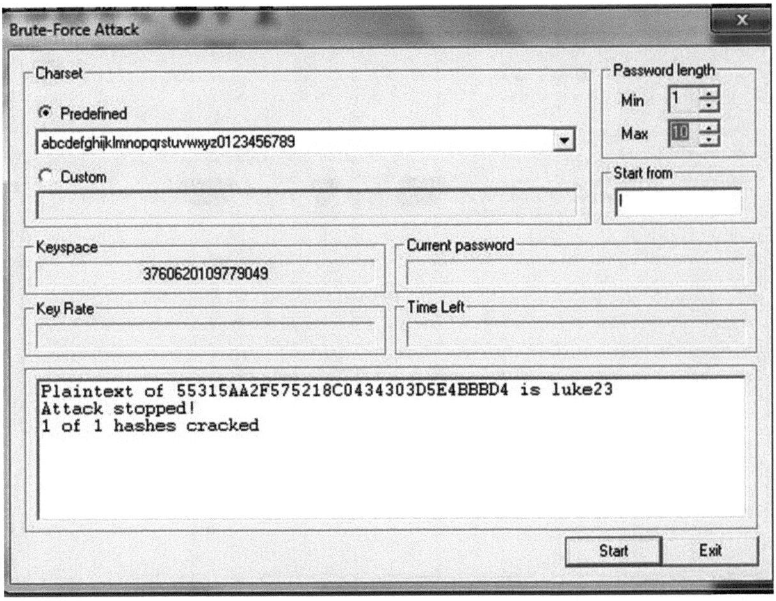

Fig. 4.9 Cain and Abel password recovery

The various attack approaches described do require certain assumptions to be in place before they become possible. For instance, in order to crack the password of a SAM file, the attacker needs to obtain a copy. In order to sniff the traffic on a network, the attacker needs to be connected to an appropriate part of the network. Hardware keyloggers will require physical access to the system so that they can be inserted online. Therefore, through additional countermeasures, it is possible to reduce the likelihood of some threats, for instance, limiting login attempts prevents active brute-force attacks.

4.3 Token Authentication

Token authentication is an approach, in concept, that has been utilised a long time prior to computers being invented. Keys to a house or a car are tokens used to verify physical access. In modern definitions, when applied to their use for electronic access, the secret information is no longer a physical key but an electronic key stored in memory within a physical device. The security they provide relies upon the owner. A lost or stolen key requires the owner to identify such loss in a timely fashion and to rectify it as quickly as possible. If this were a house key, it would require a change of locks. In the electronic world, it means a cancellation and revocation of all access rights associated with the token, the time window between loss and reporting providing an opportunity for an attacker.

Formally, the concept of token authentication is that the token is presented to the system for authentication. A conversation is had between the token and the system via some wired or wireless means and should the token have the correct password (or key as it is more generally referred to within token authentication) access is granted. The principal assumption relied upon for token-based authentication to operate securely is that the legitimate (and only the legitimate) person has possession of the token. However, no technical security mechanisms exist to verify that assumption. Merely policy-level mechanisms such as the user must report loss of tokens immediately.

Characteristics of a good token include:

- Every token must be unique.
- Tokens must be difficult to duplicate.
- Tokens must have a portable and convenient form factor.
- Tokens must be cheap and readily replaceable when lost, stolen or mislaid.

There are a wide variety of different tokens, designed to serve various purposes such as physical access to buildings, cars and offices or logical access to systems such as electronic payment, mobile phones and computers. However, broadly, tokens can be classified by the fundamental nature of their operation, the latter largely being the predecessor to the former:

- *Passive Tokens* simply store a secret, which is presented to an external system for processing and validation.
- *Active Tokens* also store a secret; however, this is never released to an external system. Instead, the token is capable of processing the secret with additional information and presents the result of the verification to an external system.

4.3.1 Passive Tokens

Passive tokens are the original form of tokens, the most familiar of which are the magnetic-strip plastic-based cards. These types of card found applications in a variety of systems including credit/debit cards, driving licenses, physical door access and national identification cards. The magnetic strip comprises three tracks, and depending upon the application, one or more of these are utilised. Within credit cards (or other financial-based cards) the information present on the magnetic strip is defined by ISO 7813 (ISO 2006), which stipulates three tracks, although track 3 is virtually unused. An illustration of the information stored on the magnetic strip is illustrated in Fig. 4.10. Track 1, recorded at a higher bit-rate, also has the capacity to include the owners' name.

The other main category of passive token is proximity-based. Proximity tokens provide authentication and subsequent access (traditionally to physical but increasingly to logical systems) by close physical proximity between a token and reader – with the reader subsequently attached to the system that you wish to be

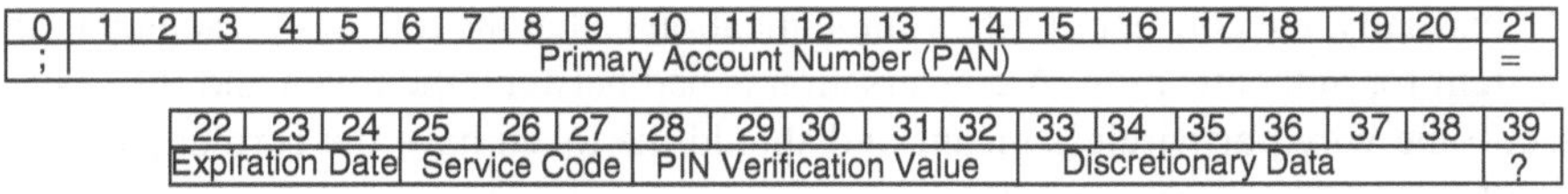

Fig. 4.10 Financial cards: Track 2 information

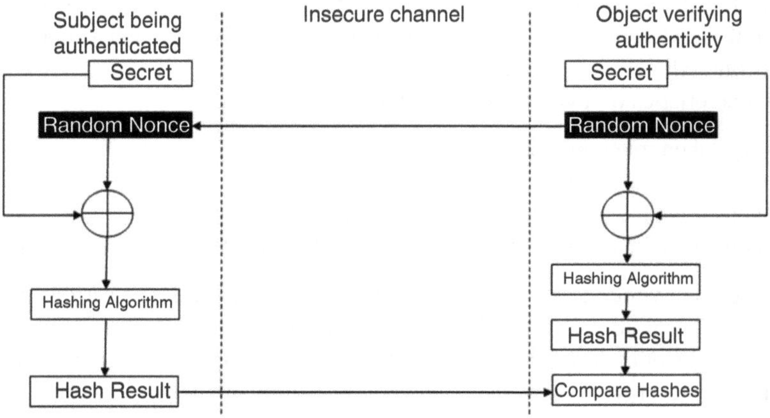

Fig. 4.11 An authentication without releasing the base-secret

authenticated upon. The distance required between the token and reader can vary from centimetres to a few meters, the former operating in a *passive* mode being powered by self-induction and without an internal power source such as a battery, and the latter referred to as *active* mode which includes a power source (hence the greater range). Obviously the choice of which mode to utilise would depend upon the security requirements of the application, with shorter-distance proximity cards providing a stronger verification of who has accessed a resource; active cards tend to be more expensive and have a shorter longevity, having a finite operational time due to the reliance upon the battery. It is worth highlighting that these proximity cards have also been referred to as radio frequency identification (RFID) cards/tokens. However, RFID technology is not confined to passive tokens and active versions also exist. A non-authentication-related application of passive RFID can be found in its use in theft detection and supply-chain management.

4.3.2 *Active Tokens*

The single largest security flaw of passive tokens is the release of the secret material. Active tokens do not present the secret knowledge to an external system, but use it in some fashion to generate another piece of information: typically a one-time password, which is subsequently presented for authentication. Figure 4.11 illustrates a simple example of a process that can be utilised to provide authentication of the token without exposing the base secret. In any authentication system, both the object requiring access and the subject of the access share the authentication credential. In order for the system to verify the authenticity of the token, it can send a random number (technically referred to as an authentication nonce) to the token. Both the nonce and the base secret are combined and then hashed (using an algorithm such as MD5 or SHA) creating a unique fingerprint of the data. The resulting hash is sent back to the system, where it is compared with its own computed hash. If they match, the token is authenticated. If you then reverse the process, with the token sending the nonce and the system returning the result of the hash, the system will have authenticated to the token. Performing both actions results in mutual authentication – an increasingly common process in many protocols.

Implemented in this manner, active tokens are immune to network sniffing and replay attacks, as the password used to obtain access is different every time. In certain scenarios, the token and server are unable to directly communicate and therefore the user is required to enter the resulting password that is based upon the base secret. As the nonce is unable to be communicated, another piece of information is utilised as a basis for the creation of the one-time password, but the key assumption in these approaches is that both the token and system *must* be synchronised so that the one-time password generated on the token is identical to the one-time password generated by the system. In order to manage this synchronisation issue, one-time password tokens primarily come in two forms: counter-based and clock-based. Counter-based tokens were the first to become commercially available and were simple in principle. Every time you needed a one-time password you pressed a button on the token and the token would generate a new password and the internal counter would increment. Every time a new password was entered into the system, the system counter was also incremented. So as long as you only pressed the button when a one-time password was required, the systems remained synchronised. Unfortunately, due to the nature of the token, often carried in pockets and bags, the button would often be pressed by mistake causing the two systems to fall out of synchronisation and subsequently not work – requiring re-synchronisation by the system administrator. To get over this user issue, clock-based tokens were created that rely upon the clocks on both systems being the same. These tokens are generally far more effective; however, time synchronisation can be more difficult to achieve over the long term due to fractional differences in clocks. The RSA SecurID is an example of a clock-based one-time password generator, as illustrated in Fig. 4.12. Whilst strictly speaking RSA SecurID is a two-factor approach combined with a secret-knowledge technique, implementations exist that simply rely upon the one-time password functionality (RSA 2011).

Despite limitations, one-time password tokens are increasing in popularity, with many banks and financial organisations providing them to their customers for increased security (particularly with regard to online banking); however, the functionality of the one-time password generated is stored within the card. The subsequent card reader is simply utilised to query the card and display the resulting one-time password (as illustrated in Fig. 4.13).

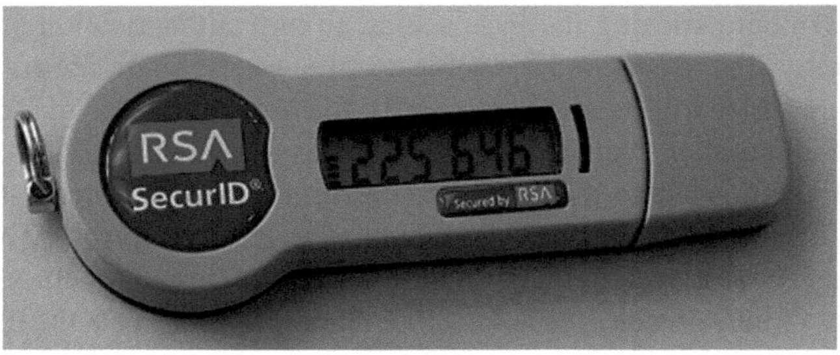

Fig. 4.12 RSA SecurID token (Source: Wikimedia Commons 2011)

Fig. 4.13 NatWest debit card and card reader

The previous example of the NatWest one-time password is effectively an example of a particular use of Smartcard technology. The most popular form of token-based authentication, it has superseded many magnetic-based implementations – most notably credit and debit cards. As the secret information remains within the card, cryptographically protected from logical access, the ability for an attacker to effectively sniff the network and utilise that information for financial benefit has been dramatically reduced. In addition, given the small form factor of the smartcard chip,

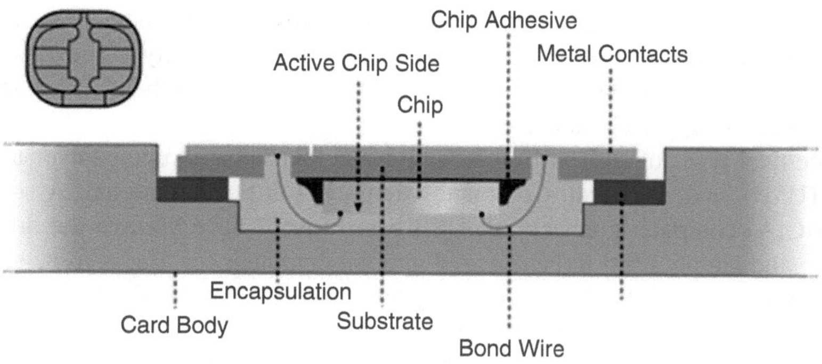

Fig. 4.14 Smartcard cross section (Source: Wikimedia Commons 2011)

physical attacks such as the ability to perform side-channel attacks (e.g. analysis of power signals within the token) have become increasingly more difficult and require a significant level of technical expertise and equipment. Many modern smartcards also now integrate tamper-resistant and tamper-evident controls within them to ensure such attacks are not possible (or at least not cost-effective to do so).

In order for Smartcards to ensure the base secret is not released they need to perform the necessary processing that would normally be undertaken by the reader (in older systems) such as a variety of cryptographic processes. To achieve this, smartcards include volatile and non-volatile memory and microprocessor components. Figure 4.14 illustrates a diagrammatic view of the cross section. The processing capabilities, cryptographic algorithms and processes it supports and the memory capacities all vary depending upon the model, with the subsequent price per unit increasing with the increasing capabilities. For large-scale implementations, the price differences can make a significant impact and therefore a trade-off exists between cost and functionality. A good example of this trade-off in practice is the London Oyster Card – a contactless smartcard principally used for authorising travel on the London Underground. Based originally upon the MIFARE Classic chip by NXP Semiconductors, the smartcard utilised propriety security algorithms, which were subsequently analysed and broken by Dutch researchers (de Winter 2008). In effect, the chip lacked any commercial-grade cryptographic processes. As such, due to the asynchronous nature of the Oyster system, attackers are able to travel on the underground for free (for a day with each compromised card) (BBC 2008). Since 2010, the Oyster card is now based upon the MIFARE DESFire chip, a cryptographically enabled smartcard with support for standardised cryptographic algorithms such as the Advanced Encryption Standard (AES) (Balaban 2010).

Contactless smartcards such as the Oyster Card are becoming a far more popular form of token. Broadly referred to as an RFID token, these contactless tokens receive their electrical power at least in part from the signal that emanates from the reader. The more processing heavy RFID chips require a secondary power supply to power the microcontroller and its supporting components, which in turn reduces the lifespan of the token. The contactless nature of the token reduces component wear-and-tear that contact-based cards and readers suffer from. However, the contactless nature of the signal and the extreme power constraints have led to a variety of security issues. Most of the issues to date, however, are focused upon non-authentication applications, such as supply-chain management.

The final types of active token are pluggable tokens based upon universal serial bus (USB) and Personal Computer Memory Card International Association (PCMCIA) computer connections. USB-based tokens have become the dominant choice and simply need to be plugged into the computer for access to be granted. The main advantage of pluggable tokens over one-time passwords is that they support more complicated authentication protocols.

The use and applicability of a token-based approach depends upon what it is being used to give access to and the level of security required to protect access. Unfortunately, tokens by themselves are not inherently good at ensuring the system gives access to the correct person rather an impostor. The system merely authenticates the presence of the token, not the person with whom it belongs. Therefore, through theft or duplication of the token, it is relatively straightforward for an attacker to compromise a system.

It is therefore not unsurprising that tokens are most commonly utilised in a two-factor authentication mode. The use of the token–password approach is very well established, with many commercial companies deploying large-scale systems. Credit cards and debit cards are excellent examples, where the card is paired with a PIN. The token–biometric approach is newer and increasingly been used in logical access control, e.g. as a pluggable USB key with integrated fingerprint reader. Although the use of two-factor authentication increases the overall security by providing an additional layer of complexity, it is important to realise that these systems are not infallible.

4.3.3 Attacks Against Tokens

A major threat to tokens is theft of the token itself. In systems that rely only upon the token, security is effectively breached the instant the theft takes place. Although strong procedural policies can be put in place by organisations to ensure lost or stolen cards are reported and access prohibited, it would be ill conceived to rely upon the users to provide effective access control.

Rather than stealing the token, attackers seek to duplicate it, with the likelihood that such misuse would go unnoticed by the owner for a longer period of time, permitting a greater degree of misuse. The main problem with passive tokens is the increasingly simple ability to duplicate the contents of the token. For instance, whilst the technology involved in magnetic-based credit cards used to be quite expensive and thus a barrier for some less motivated attackers, technology advancements have drastically reduced this. The equipment required to copy or *skim* a magnetic-stripe card costs less than $10. In 2004, prior to Chip and PIN, the UK Payments Association (formally APACS) reported a fraudulent card transaction

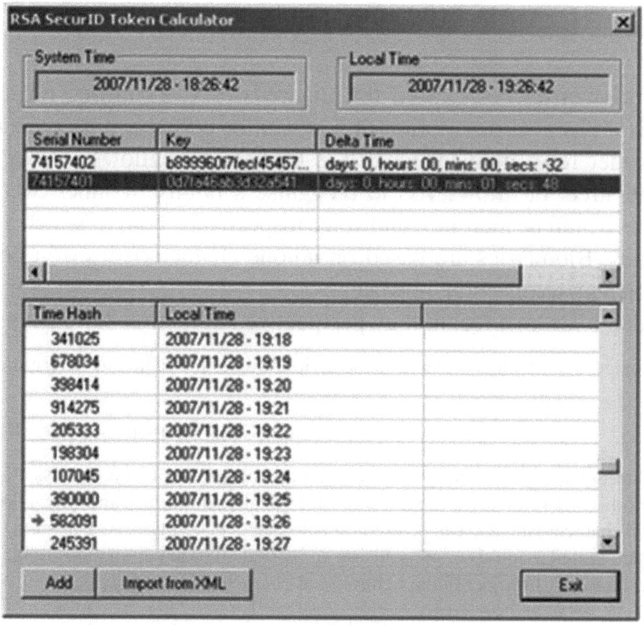

Fig. 4.15 Cain and Abel's RSA SecurID token calculator

every 8s (Chip and PIN 2006). The addition of Chip and PIN and two-factor authentication on bank cards for purchases (as well as for use in an ATM) has assisted in reducing fraud; however, the lack of a PIN for Internet-based purchases has simply moved the fraud from the store to online. Credit card companies are addressing this issue through schemes such as *Verified by Visa* and *MasterCard SecureCode*, which require additional secret knowledge–based verification.

Although active tokens have provided an additional layer of protection, there are a number of attacks that exist against them. For example: IP hijacking of sessions by sniffing network traffic whilst the one-time passwords is transmitted; man in the middle attacks and side-channel analysis of smartcards by power consumption, to name but a few. Cain and Abel includes an RSA SecureID token calculator that is able to replicate the functionality of the token (as illustrated in Fig. 4.15). It requires two key pieces of information: the serial number and the token activation key, but once configured is able to provide all the one-time passwords. This information is usually provided to the legitimate user in XML files – so access to these would be required.

Additional countermeasures, such as tamper-evidence, are being included with Smartcards to make the attack more challenging. Compared to the simple theft of the token these attacks are considerably more technically challenging, effectively raising the 'technological bar' required by attackers to compromise the system.

4.4 Biometric Authentication

The use of biometrics has existed for hundreds of years in one form or another, whether it is a physical description of a person or perhaps more recently a photograph. Consider for a moment what it is that actually allows you to recognise a friend in the street or allows you to recognise a family member over the phone. Typically this would be their face and voice respectively, both of which are biometric characteristics. Biometrics are based on unique characteristics of a person – how unique is often up to question, however.

More recently, from a technology perspective, biometrics have been defined as

> 'the automated use of physiological or behavioural characteristics to determine or verify identity'
>
> International Biometrics Group (IBG 2010a).

> 'a general term used alternatively to describe a characteristic or as a process. As a characteristic [biometrics refer to] a measurable biological (anatomical and physiological) and behavioural characteristic that can be used for automated recognition. As a process [biometrics refers to] automated methods of recognising an individual based upon measurable biological (anatomical and physiological) and behavioural characteristics'.
>
> National Science and Technology Council (NSTC 2006).

No single definition for the term exists to date, although a standard is under development by the International Organization for Standardization (ISO) Joint Technical Committee (JTC) 1 Special Committee 37 on a harmonised biometric vocabulary (ISO 2010). However, from the above definitions, some particular aspects of the definition are similar. Terms that seem pertinent are automated, physiological, behavioural, determine and verify. The automated process of sample capture, process, extraction, classification and decision defines a biometric system over simply a possible biometric characteristic. For instance, whilst DNA certainly has the property of uniqueness to be used as a biometric, the process is not automated and requires human intervention. It therefore, formally, is not a biometric – at present! Physiological and behavioural refer to the two categories of biometric approaches, each with their own set of properties. The former refers to characteristics that are physical in nature, many of which are largely determined before birth, for instance, fingerprints and the way we look. The latter refers to characteristics that have been learnt or are dependent upon the environment with which the person has lived, for instance, the way in which people walk or sign their name. Some characteristics contain aspects of both categories, such as a person's voice. The decision regarding which category a technique belongs to is made based upon the one that has the more significant contribution to make. For instance, the characteristics that make up the voice are dependent upon physical aspects such as the vocal tract and mouth. However, it is widely accepted that many of the unique characteristics of the voice, such as use of language and accent, are developed from learnt behaviour. Finally, the definition refers to determining or verifying identity. This aspect of the definition highlights the two modes in which biometric systems can operate:

- Verification – determining whether a person is who they claim to be.
- Identification[2] – determining who the person is.

The particular choice of biometric will greatly depend upon which of these two methods is required, as performance, usability, privacy and cost will vary accordingly. Verification, from a classification perspective, is the simpler of the two methods, as it requires a one-to-one comparison between a recently captured sample and a reference sample, known as a template, of the claimed individual. Identification requires a sample to be compared against every reference sample, a one-to-many comparison, contained within a database, in order to find if a match exists. Therefore the unique characteristics used in discriminating people need to be more distinct or unique for identification than for verification. Biometrics are not based upon completely unique characteristics. Instead a compromise exists between the level of security required and thus more discriminating characteristics and the complexity, intrusiveness and cost of the system to deploy. It is unlikely, however, in the majority of situations that a choice would exist between which method(s) to implement. Instead, different applications or scenarios tend to lend themselves to a particular method. For instance, PC login access is typically a verification task, as the user will select their username. However, when it comes to a scenario such as claiming benefits, an identification system is necessary to ensure that the person has not previously claimed benefits under a pseudonym. Whilst identification will be referred to occasionally in this book, the primary focus is placed upon verification. As such, unless stated otherwise, assume the process of verification is being referred to.

4.4.1 Biometric System

A biometric system is a significantly more challenging authentication approach to design than passwords and tokens. With both secret-knowledge and token-based approaches, effectively the process entails comparing two binary sequences and making a decision to permit or deny access based upon whether they are identical or not. Biometrics comprise of a number of components, whose complexity will depend upon the individual set of biometric characteristics being used. The components are listed in Table 4.5 and the process is illustrated in Fig. 4.16.

[2] Literature also classifies identification in two further modes: open-set and closed-set identification. Open-set identification refers to identifying if someone is in the database and if so finding the record. In closed-set identification, it is assumed that the person is in the database, and the system needs to find the correct record. Whilst they appear similar in operation, the slight difference in assumptions of whether the individual is in the database or not results in a significant difference in system complexity, with the open-set identification being a far more challenging system to develop.

Table 4.5 Components of a biometric system

Component	Description
Capture	The process that obtains the biometric sample from a user. This will involve some form of sensor technology, which can take many forms depending upon the biometric approach, from a simple web camera for facial recognition to optical sensors for fingerprint recognition
Extraction	The extraction phase takes the captured image or sample and extracts the unique biometric information from the sample. For many biometric techniques this can be a number of measurements taken at particular points in the sample. For instance, in facial recognition, the capture phase takes a photograph of the individual and the extraction phase will take a number of measurements of the face that uniquely describe the facial characteristics. The resulting output from the extraction phase is referred to as a feature vector
Storage	The storage element merely permits storage of the feature vector for use in subsequent comparisons. For identification systems, the storage component is far more sophisticated than for verification to allow for fast indexing and searching of the database
Classification	The classification phase is a process that compares two samples: the original sample taken initially at setup and a new sample. The output of this process is a value that indicates the level of similarity between the two samples. Typically, this can be a value between 0 and 1 – 1 indicating a perfect match
Decision	The final component is the decision phase. It has the responsibility for determining whether to grant or deny access. To make the decision a threshold is used. Should the sample meet or exceed the threshold, the sample is approved; should the sample similarity score be less than the threshold, it is rejected. In addition to the output from the classification phase, additional information can also be used in the decision-making process, such as the quality of the sample provided or soft biometric information such as gender

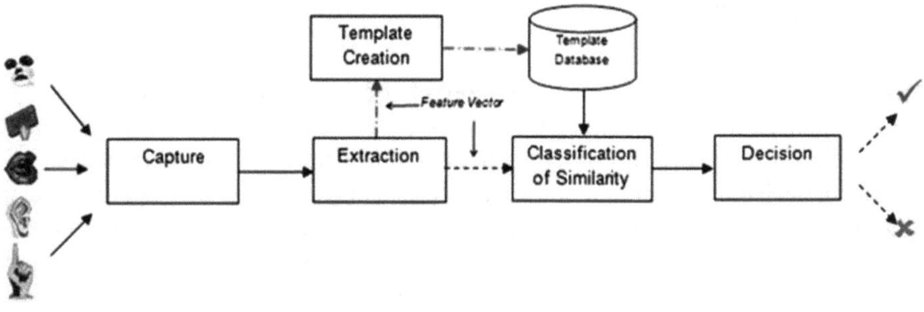

Fig. 4.16 The biometric process

The key processes involved in biometric systems are enrolment and authentication. Enrolment describes the process by which a user's biometric sample is initially taken and used to create a reference template for use in subsequent authentication. As such, it is imperative that the sample taken during enrolment is from the authorised user and not an impostor, and that the quality of the sample is good. The actual number of samples required to generate an enrolment template will vary according to the technique and the user. Typically, the enrolment stage will include a quality check to ensure the template is of sufficient quality to be used. In cases where it is not, the user is requested to re-enrol onto the system.

Authentication is the process that describes the comparison of an input sample against one or more reference samples – one in the context of a verification system, many with an identification system. The process begins with the capture of a biometric sample, often from a specialised sensor. The biometric or discriminatory information is extracted from the sample, removing the erroneous data. The sample is then compared against the reference template. This comparison performs a correlation between the two samples and generates a measure or probability of similarity. The threshold level controls the decision as to whether the sample is valid or not, by determining the required level of correlation between the samples. This is an important consideration in the design of a biometric, as even with strong biometric techniques, a poorly selected threshold level can compromise the security provided. Finally, the decision is typically passed to a policy management system, which has control over a user's access privileges.

In practice, the realisation of biometric systems can take many forms depending upon the objectives and scale of the implementation. For instance, it is possible to have a biometric system where all components are highly integrated into a single device; capture, extraction, storage and classification are all performed locally. This has the advantage that the device is less prone to various attacks as the sensitive biometric data never leaves the device. If the sensor is a one-off device used to protect access to a highly secure server room to which few people have access, or perhaps for personal use on a mobile phone, this solution is effective. However, should you wish to use biometrics for physical access control, where the building has many entrances and many employees, a network-based solution is necessary where authentication is performed and monitored by a central server. If this were not the case, the system administrator would be responsible for ensuring each of the biometric devices are loaded with the necessary biometric templates. This would need updating continuously to reflect changes in employees and prevents any central monitoring of the system. However, network-based solutions also open the door to various network-based attacks. It is important therefore to ensure an appropriate architecture is developed to meet the specific requirements of the deployment.

The selection of which biometric technique to utilise is not a simple decision to make. Unfortunately, whilst biometrics rely on uniqueness, no biometric technique is 100% unique, even if the characteristic were to have the property, as the system will add errors through noise in the sample capture and extraction processes. Uniqueness, however, is not the only criterion used to decide which technique to implement. Several other characteristics or factors are utilised as illustrated in Table 4.6.

Each biometric technique has a different constitution of properties. Retina and iris recognition are amongst the most unique approaches with very time-invariant or permanent features. They are also the most challenging to collect and invariably have issues with acceptability of users due to the difficulty in capturing a good-quality sample. In general, physiological biometrics tend to perform better on

Table 4.6 Attributes of a biometric approach

Attribute	Description
Uniqueness	The ability to successfully discriminate people. More unique features will enable more successful discrimination of a user from a larger population than techniques with less distinctiveness
Universal	The ability for a technique to be applied to a whole population of users. Do all users have the characteristics required? For instance, users without fingerprints.
Permanence	The ability for the characteristics not to change with time. An approach where the characteristics change with time will require more frequent updating of the biometric template and result in increased cost to maintenance
Collectable	The ease with which a sensor is able to collect the sample. Does the technique require physical contact with the sensor, or can the sample be taken with little or no explicit interaction from a person. What happens when a user has broken their arm and is subsequently unable to present their finger or hand to the system?
Acceptable	The degree to which the technique is found to be acceptable by a person. Is the technique too invasive? Techniques not acceptable will experience poor adoption and high levels of circumvention and abuse
Circumventable/Unforgeablity	The ability not to duplicate or copy a sample. Approaches that utilise additional characteristics, such as liveness testing, can improve the protection against forging the sample

uniqueness and permanence and behavioural biometrics are easier to collect and have better levels of acceptability. With regard to how universal a biometric is, this will tend to vary on a technique-by-technique basis. All biometrics have a fairly high degree of universality, but there is always a population of people that are unable to successfully provide a sample. For instance, techniques that rely upon a limb being present, such as fingerprint, hand geometry and vascular pattern recognition, would not be possible for individuals missing those particular limbs. Facial recognition is possibly the most universal of the techniques; however, it can suffer from low levels of permanence, with people's faces changing in relatively short periods of time due to weight gain or loss or perhaps hair growth. The degree to which each technique is susceptible to circumvention through spoofing (the creation of a fraudulent sample) does vary between techniques. However, those approaches that obtain samples from within the body rather on the surface are far more challenging to spoof. For instance, vascular pattern recognition and retina scanning are both very difficult to spoof. Fingerprint and facial recognition are less so, as the process of copying the trait is simpler.

The level of entropy that exists within biometrics is not fixed at the biometric level; nor is it fixed at the individual approach level. It is wholly dependent upon the feature vector produced (which is intrinsically coupled to the method of classification). Even within a biometric technique, the entropy achieved can vary considerably.

Moreover, with many proprietary classification approaches it is difficult to determine the actual level of entropy. This situation is further complicated by the fact that within biometric systems an exact match between samples is not required – merely a sufficient level of similarity, with sufficiency being determined by the threshold level, which is defined on a system-by-system basis. It is therefore not easy to determine the precise entropy levels for biometric systems for a direct comparison. It is, however, generally understood that the more unique biometrics have extremely high levels of entropy. For instance, iris recognition can have up to 400 data points, with each data point consisting of a large range of values (for illustrative purposes say 100), which provides an entropy of 865 bits.

4.4.2 Biometric Performance Metrics

Given all biometrics work on the basis of comparing a biometric sample against a known template, which is securely acquired from the user when they are initially enrolled on the system, the template-matching process gives rise to misclassifications, of both the authorised user and impostors. In verification systems, these misclassifications result in a characteristic performance plot between the two main errors governing biometrics, which are:

- *False acceptance rate* (*FAR*), or rate at which an impostor is accepted by the system, and
- *False rejection rate* (*FRR*), or rate at which the authorised user is rejected from the system.

The error rates share a mutually exclusive relationship – as one error rate decreases, the other tends to increase, giving rise to a situation where neither of the error rates is typically at 0%. Figure 4.17 illustrates an example of this relationship. This mutually exclusive relationship translates into a trade-off for the system designer between high security and low user convenience (tight-threshold setting) or low security and high user convenience (slack-threshold setting). Typically, a threshold setting that meets the joint requirements of security and user convenience is usually set. A third error rate, the *equal error rate* (*EER*), equates to the point at which the FAR and FRR meet and is typically used as a means of comparing the performance of different biometric techniques. These performance rates when presented are the averaged results across a test population, therefore presenting the typical performance a user might expect to achieve. Individual performances will tend to fluctuate depending upon the uniqueness of the particular sample.

The FAR and FRR metrics refer to the overall system performance of the biometric as determined at the decision stage. These rates are also accompanied by:

- *True acceptance rate* (*TAR*), the rate at which the system correctly verifies the claimed individual, and
- *True rejection rate* (*TRR*), the rate at which the system correctly rejects a false claim.

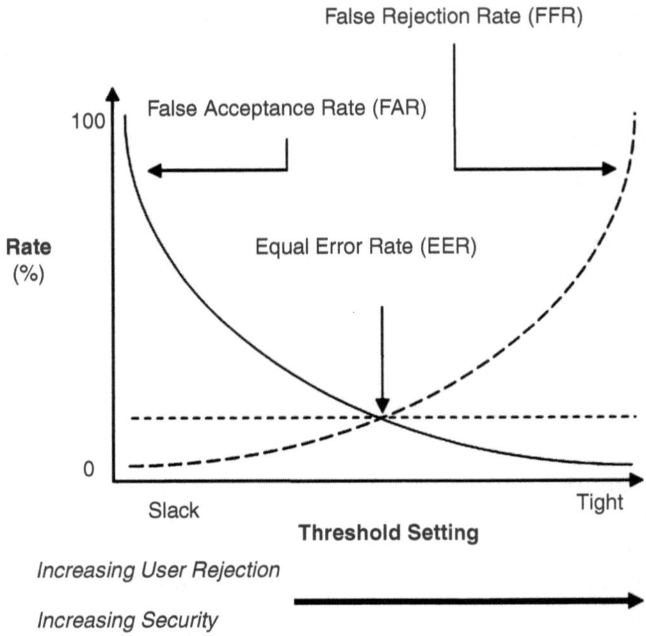

Fig. 4.17 FAR/FRR performance curves

It is more common for biometric vendors to publish their rates in terms of the FAR and FRR. There are, however, a number of other metrics commonly used when testing and evaluating biometric systems. These are calculated at extraction and matching stages of the biometric process. At the extraction phase the following metrics are established:

- *Failure to acquire (FTA)*, the rate at which either the capture or extraction stage is unable to create a valid template, and
- *Failure to enrol (FTE)*, the rate at which the user is unable to successfully enrol onto the system.

Reasons for failure are numerous but can include problems with the physical sample such as dirty or scarred fingerprints; problems with the sensor such as swiping the finger too quickly and environment factors such as poor illumination for facial recognition. Both error rates are typically associated with problems capturing the sample due to poor human–computer interaction (or a lack of education on how to use the system) and the sensitivity of the capture device. The FTA is a contributing factor of the FTE as it measures the proportion of the population that does not have a particular trait or is unable to provide one of sufficient quality. FTA errors can also appear for individuals who have successfully enrolled. As such the FRR and FAR metrics include the FTA errors. The FAR metric also includes the FTE. The FTE is not included within the FRR because if the user is never enrolled, they are unable to be falsely rejected.

The remaining error rates are associated with the matching or classification stage. These mirror the FRR and FAR and are referred to as:

- *False match rate (FMR)*, the rate at which impostors are falsely verified against the claimed identity, and
- *False non-match rate (FNMR)*, the rate at which authorised users are incorrectly rejected.

The key difference between the FMR/FNMR and the FAR/FRR metrics is that the latter includes other errors that appear in the system – for instance, the FAR includes the FMR, FTA and FTE errors, and therefore presents the overall system performance.

The metrics are related as follows:

$$\mathrm{FNMR} = 1 - \mathrm{TAR}$$
$$\mathrm{FMR} = 1 - \mathrm{TRR}$$

$$\mathrm{FRR} = \mathrm{FNMR} + \mathrm{FTA} + \textit{Decision Error}$$
$$\mathrm{FAR} = \mathrm{FMR} + \mathrm{FTA} + \mathrm{FTE} + \textit{Decision Error}$$

It should be noted, however, that both the FRR and FAR metrics can also include an error generated by the decision phase should it introduce additional information such as sample quality scores into the decision-making process. Whilst the FAR, FRR, FMR and FNMR have specific definitions, it is common in the literature for these terms to be interchanged. Care should therefore be taken in interpreting the results correctly.

There are also metrics utilised when the system operates in identification. For completion these are highlighted below:

- True positive identification rate (TPIR), the rate at which the users' correct identifier is among those returned by the system. In identification systems, a ranked list of users mostly closely associated with the profile can be returned instead of only one user. It is assumed the user is enrolled in the system.
- False positive identification rate (FPIR), the rate at which users not enrolled onto the system are falsely identified. This error rate is only present in open-set identification, as in closed-set identification all users are enrolled onto the system.
- False negative identification rate (FNIR), the rate at which users' correct identifier is not among those returned by the system. It is assumed the user is enrolled and is directly related to the TPIR (FNIR = 1 – TPIR).

The actual performance of different biometric techniques varies considerably with the uniqueness and sophistication of the pattern classification engine. In addition, published performances from companies often portray a better performance than typically achieved given the tightly controlled conditions in which they perform their test. Care must be taken when interpreting results to ensure they meet the require levels of performance for the system.

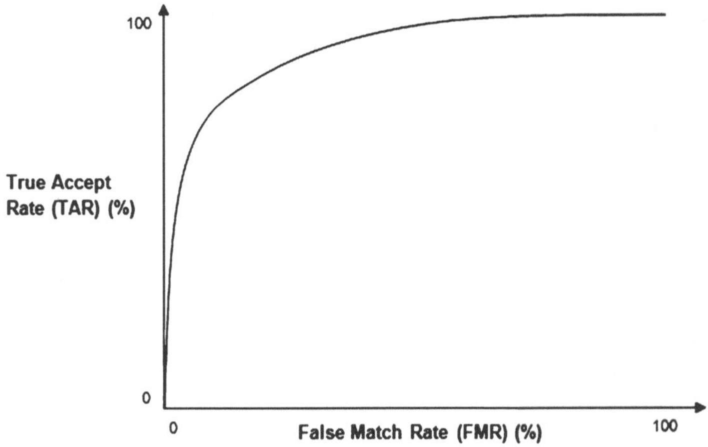

Fig. 4.18 ROC curve (TAR against FMR)

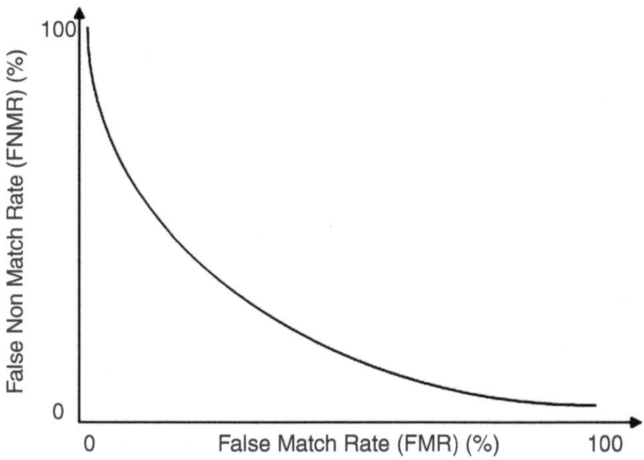

Fig. 4.19 ROC curve (FNMR against FMR)

In addition to the performance metrics, a number of curves are also utilised to graphically assist the designer in understanding the relationship between metrics and provide a mechanism for selecting the optimum threshold. Figure 4.17 illustrates the relationship between the FAR and FRR against varying thresholds. This provides a mechanism for establishing the threshold given acceptable levels of FAR and FRR error. The receiver operating characteristic (ROC) curve is also a standard graph that relates the FNMR with the TAR (or FMR). Examples of the ROC are illustrated in Figs. 4.18 and 4.19. These charts map the error rates as a function of the threshold and enable a simpler mechanism for comparing biometric techniques or algorithms.

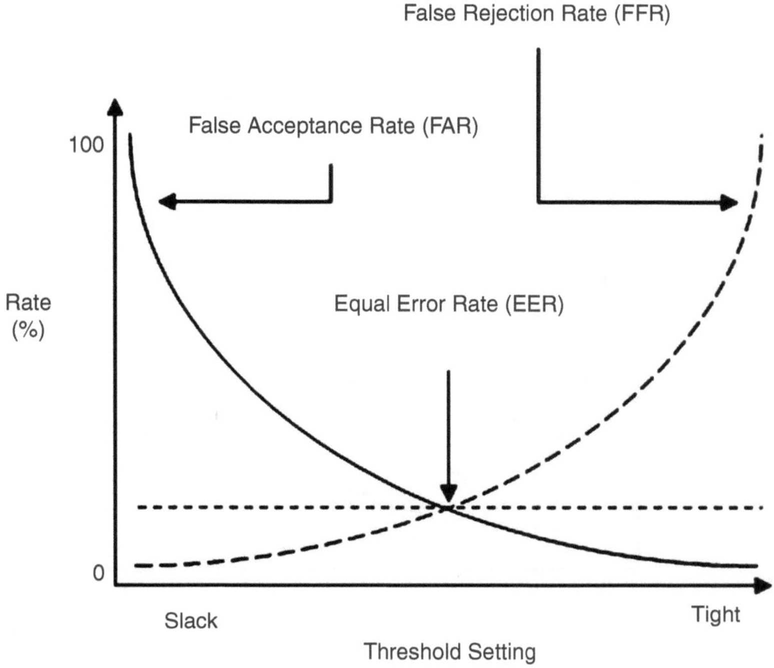

Fig. 4.20 Characteristic FAR/FRR performance plot versus threshold

Determining an appropriate threshold between performance metrics is no simple task. Unfortunately, the graphs depicting system performance and trade-off merely present the averaged performance across a test-population and hide the underlying characteristics of individual levels of performance. The reality of the system is that the FAR and FRR convey very uncharacteristic relationships which can vary significantly between users. To illustrate this point, let's say the threshold level had been set at the point at which the FAR and FRR meet, referred to as the equal error rate (EER) in Fig. 4.20, which the system designer had deemed appropriate given the trade-off between security and user convenience.

If individual users' characteristic plots are then considered, illustrated in Figs. 4.21 and 4.22, it can be seen that the previous threshold setting is not ideal for either user. For the user displaying the characteristics in Fig. 4.21, this choice of threshold level will result in a high level of user inconvenience and a higher level of security than was deemed appropriate by the system designer. For the user in Fig. 4.22, the level of security provided will be far lower than the system designer had set. So how does this threshold level get set in practice? There are only two choices, you either set a system-wide setting where all authentications are compared against the same level for all users, or set individual threshold levels for all users (with the latter obviously providing a more optimised configuration than the former). Given appropriate risk analysis and knowledge of the performance characteristics it

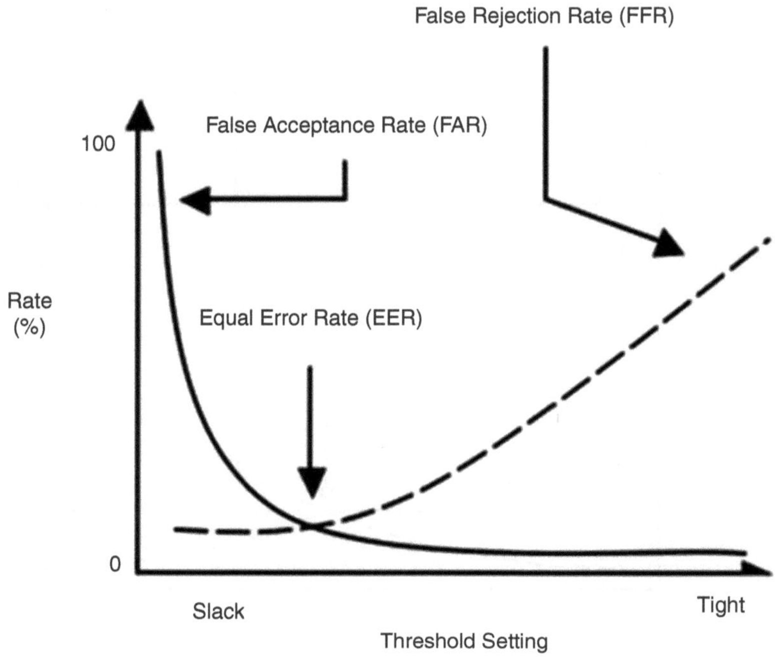

Fig. 4.21 User A performance characteristics

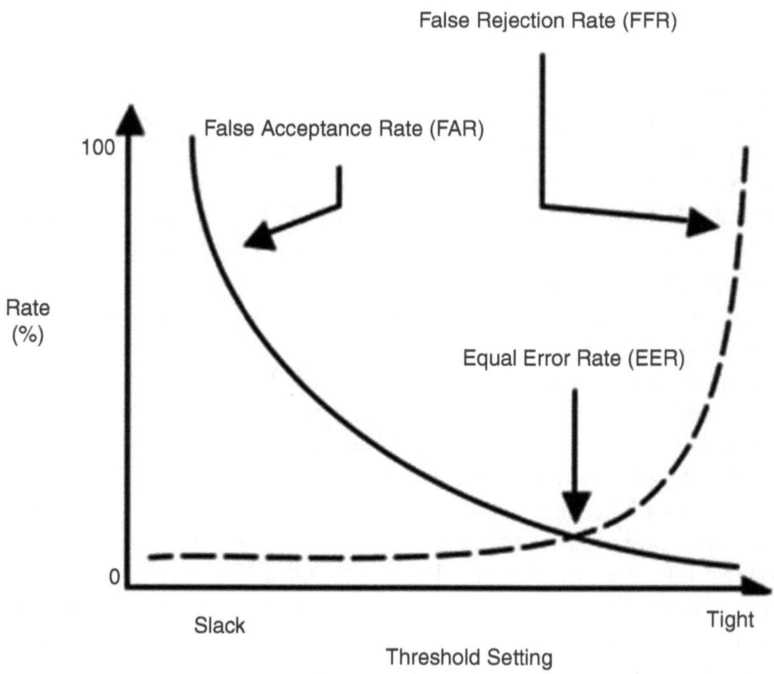

Fig. 4.22 User B performance characteristics

would be possible to define a system-wide threshold level that is appropriate to meet the security requirements of the system given a defined level of user inconvenience and a degree of tolerance. Setting such a level on an individual basis is a far larger problem, in terms of the time taken to set such a parameter, and who will have the authority to set it. Remembering it is the threshold level that ends up being the key to the biometric system, a poorly selected threshold level can remove any security the biometric technique is deemed to have.

Given this problem, time and effort have been put into finding methods of normalising the output of the pattern classification process – so that an output value of say 0.6 means the same across a population of users. Other efforts have gone into finding methods of automating the threshold decision based on a number of authorised and impostor samples – determining the performance of the biometric technique for each and every user prior to operation.

4.4.3 Physiological Biometric Approaches

The majority of core biometric techniques commercially available are physiologically based and tend to have a more mature and proven technology. In addition, physiological biometrics typically have more discriminative and characteristic invariant features, and as such are often utilised in both verification and identification systems (Woodward et al. 2003). This section will briefly provide a background to the following physiological approaches:

- Ear geometry
- Facial recognition
- Facial thermogram
- Fingerprint recognition
- Hand geometry
- Iris recognition
- Retinal recognition
- Vascular pattern recognition

It is outside the scope of this book to describe the individual biometrics in detail, but rather an overview of their characteristics and operation is provided. However, Chap. 5 will provide a detailed description of biometrics pertaining to transparency and furthermore a number of excellent texts already exist that provide this generic coverage – in particular, the reader should consider consulting the following texts if more information is required:

- Handbook of Biometrics, edited by Jain, Flynn and Ross and published by Springer (2008)
- Biometrics: Identity Verification in a Networked World, by Nanavati, Thieme and Nanavati. Published by John Wiley and Sons (2002)
- Biometrics: Identity Assurance in the Information Age by Wordward, Orlans and Higgins. Published by McGraw-Hill Osborne (2003)

Please take care when considering the performance of the biometric approaches. Many of the statistics presented are based upon studies undertaken by researchers

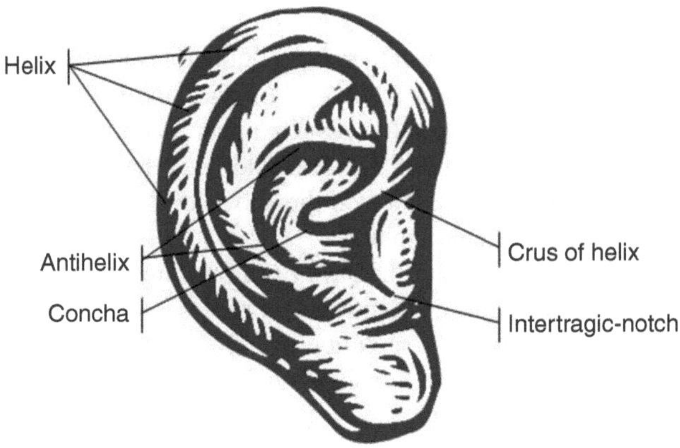

Fig. 4.23 Anatomy of the ear

and industry bodies that vary in the methodology deployed. The results are therefore not directly comparable but do provide a basis for appreciating the current level of recognition performance that can be achieved.

4.4.3.1 Ear Geometry

The human ear is a structure that contains a rich canvas of information that changes little over time. As illustrated in Fig. 4.23, the ear is made up of several components, including the helix, antihelix, concha and intertragic-notch, which results in a fairly unique pattern. Studies have reported recognition rates between 93% and 99.6% under (specific) experimental conditions (Moreno and Sanchez 1999; Hurley et al. 2005).

It is also an approach that performs well against the criterion of being universal in nature – few people are born without an ear. Understanding its acceptability is difficult to establish, as no commercial biometric systems implementing ear geometry currently exist for user authentication. The approach has almost exclusively been used in the domain of law enforcement and forensics. Whilst collecting the sample might have its difficulties, studies have argued that ear geometry does not suffer so badly from environmental factors such as facial recognition or from facial movements or orientation issues and is sufficiently large to capture from a distance unlike iris and retina recognition approaches (Jain et al. 2008).

4.4.3.2 Facial Recognition

Utilising the distinctive features of a face, facial recognition has found increasing popularity in both computer/access security and crowd surveillance applications, due in part to the increasing performance of the more recent algorithms and its covert nature (i.e. authentication of the user can happen without their explicit interaction with a device or sensor). Whilst the approach performs well against the criteria of universality and acceptability (in most scenarios), it is an approach that suffers from issues surrounding permanence and collectability. Obviously over longer periods of time (decades) the nature of the face and subsequently the facial features can change dramatically. However, shorter-term changes such as weight loss or gain can also affect the performance of the system. Facial characteristics of the very young are also likely to have not matured and be subject to change over shorter periods of time. Collectability issues can result from objects obscuring the capture (e.g. beards, glasses and hats) or the environment (e.g. illumination, position of face and distance from camera). The actual features utilised tend to change between proprietary algorithms but include measurements that tend not to change over time, such as the distance between the eyes and nose, areas around cheekbones and the sides of the mouth.

The performance of facial recognition algorithms can vary considerably depending upon the context (e.g. high-resolution images versus low-resolution, position and environment) and the extraction and classification algorithms employed. The National Institute of Standards and Technology's (NIST) large-scale experimental results from the Face Recognition Vendor Test put the approximate FAR at 0.001% with a FRR of 0.01% (Phillips et al. 2007). Whilst front-facing two-dimensional (2D) images were traditionally used in the capturing process, more recent research has focused upon three-dimensional (3D) imagery. A key advantage of the 3D image is that it is capable of authenticating users with an angle of up to 90°, rather than up to 20° with traditional 2D images.

4.4.3.3 Facial Thermogram

Facial thermogram is an approach that has evolved from the facial recognition domain and the problem of sufficient illumination for the camera to take a quality picture. The need for sufficient light restricts the applicability of facial recognition in a number of application scenarios where such control over light is not possible. Facial thermogram utilises an infrared camera to capture the heat pattern of a face caused by the blood flow under the skin. The uniqueness is present through the vein and tissue structure of a user's face. The use of an infrared camera removes the necessity for any illumination (although in practice many studies have also used intensified near-infrared sources which require some ambient light). Recent studies have shown that the external factors such as surrounding temperature play an important role in the performance of the recognition (Socolinsky and Selinger 2004). Studies have also demonstrated that the performance achieved is between 84% and 93% (Socolinsky and Selinger 2003). However, the majority of research has combined thermogram with visible sensors to augment standard recognition performance, suggesting limited scope for the approach in a unimodal approach.

Fig. 4.24 Fingerprint sensor devices

4.4.3.4 Fingerprint Recognition

Fingerprint recognition is the most widely utilised technique, with obvious applications in law enforcement as well as computer systems. They can utilise a number of approaches to classification including minutiae-based (irregularities within fingerprint ridges), ridge-based and correlation-based (Maltoni et al. 2005). The 2006 Fingerprint Verification Competition (FVC) best result obtained an average EER of 2.155% for one of the algorithms (FVC2006 2006). The image capture process does require specialised hardware, based upon one of four core techniques: capacitive, optical, thermal and ultrasound, with each device producing an image of the fingerprint. Figure 4.24 illustrates the more common optical and capacitive scanners, the latter comprising a smaller form-factor than the former but frequently resulting in a poorer image.

Fingerprint recognition is a mature and proven technology with very solid and time-invariant discriminative features suitable for identification systems. Although the uniqueness of fingerprints is not in question, with identical twins even having different prints, fingerprint systems do suffer from usability problems such as fingerprint placement, dirt and small cuts on the finger. To date, fingerprint recognition has been deployed in a wide variety of scenarios from access security to computer security on laptops, mobile phones and personal digital assistants (PDAs). Acceptability overall is quite good, as people have a better understanding of the technique – largely derived from its use within law enforcement.

4.4.3.5 Hand Geometry

Hand geometry involves the use of a specialist scanner, which takes a number of measurements such as length, width, thickness and surface area of the fingers and hand (Smith 2002). Different proprietary systems take differing numbers of

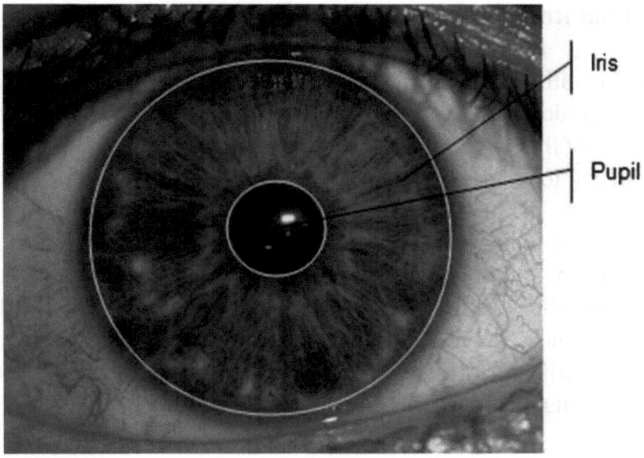

Fig. 4.25 Anatomy of an iris (Modified from original: Wikimedia Commons 2011)

measurements but all the systems are loosely based on the same set of characteristics. Unfortunately, these characteristics do not tend to be unique enough for large-scale identification systems, but are often used for time and attendance systems (Ashbourn 2000). The sensor and hardware required to capture the image tends to be relatively large and arguably not suitable for many applications such as computer-based login (Ingersoll-Rand 2011).

4.4.3.6 Iris Recognition

The iris is the coloured tissue surrounding the pupil of the eye and is composed of intricate patterns with many furrows and ridges, as illustrated in Fig. 4.25. The iris is an ideal biometric in terms of both its uniqueness and stability (variation with time), with extremely fast and accurate results (Daugman 1994).

Traditionally, systems required a very short focal length for capturing the image (e.g. physical access systems), increasing the intrusiveness of the approach. However, newer desktop-based systems for logical access are acquiring images at distances up to 40 cm (Nanavati et al. 2003). Cameras are still, however, sensitive to eye alignment, causing inconvenience to users. In terms of performance, Daugman reports a best EER of 0.0011 from the NIST Iris Competition Evaluation (ICE) (Jain et al. 2008). Iris recognition is therefore suited to identification scenarios and this is where the majority of implementations to date have been deployed. For example, the UK utilised iris recognition for expediting passport checks at airports and the UAE has integrated iris recognition into all ingress points (Crown Copyright 2010; IrsisGuard 2004).

4.4.3.7 Retinal Recognition

Retina scanning utilises the distinctive characteristics of the retina and can be deployed in both identification and verification modes. An infrared camera is used to take a picture of the retina, highlighting the unique pattern of veins at the back of the eye. Similarly to iris recognition, this technique suffers from the problems of user inconvenience, intrusiveness and limited application as the person is required to carefully present their eyes to the camera at very close proximity. In addition, hardware has been traditionally prohibitively expensive. As such, the technique tends to be most often deployed within physical access solutions with very high security requirements (Nanavati et al. 2002). The approach does have excellent performance characteristics with the extraction phase producing upward of 400 data points (30–40 points for fingerprint minutiae) (Das 2007).

4.4.3.8 Vascular Pattern Recognition

This technique utilises the subcutaneous vascular network on the back of the hand to verify an individual's identity. The patterns are highly distinctive and different in identical twins. The subcutaneous nature of the technique, like retinal recognition, is difficult to falsify or spoof. Vascular pattern recognition does not suffer from the same usability issues, however. As such, this technique has experienced significant focus with commercial products available.

The requirement to capture the hand results in a fairly large sensor, which in turn makes it an inappropriate technology for computer access (in the most part). Applications of the technology include physical access and specific scenarios such as automated teller machine (ATM) authentication, where the technology has the opportunity to be developed within the system itself. The performance of the approach, as reported by studies, suggests an EER in the region of 0.145% (with some reporting even better performances) (Miura et al. 2004).

Additional physiological biometrics have been proposed such as odour and fingernail bed recognition, with research continuing to identify body parts and other areas with possible biometric applications (Woodward et al. 2003).

4.4.4 Behavioural Biometric Approaches

Behavioural biometrics classify a person on some unique behaviour. However, as behaviours tend to change over time due for instance to environmental, societal and health variations, the discriminating characteristics used in recognition also change. This is not necessarily a major issue if the behavioural biometric has built-in countermeasures that constantly monitor the reference template and new samples to ensure its continued validity over time, without compromising the security of the technique. In general, behavioural biometrics tend to be more transparent and

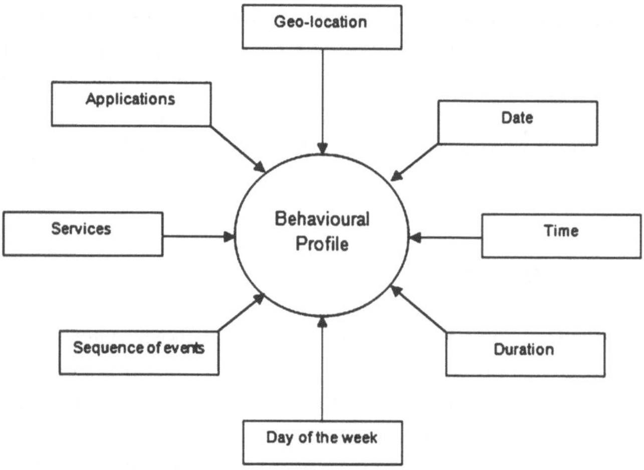

Fig. 4.26 Attributes of behavioural profiling

user-convenient than their physiological counterparts, however, at the expense of a lower authentication performance.

This section will briefly explain the following behavioural approaches to biometrics:

- Behavioural profiling
- Gait recognition
- Keystroke analysis
- Signature recognition
- Speaker recognition

4.4.4.1 Behavioural Profiling

Behavioural profiling (also referred to as service utilisation) describes the process of authenticating a person based upon their specific interactions with applications and/or services. For instance, within a PC, service utilisation would determine the authenticity of the person depending upon which applications they used, when and for how long, in addition to also utilising other factors. The permanence of the features is poor with the variance experienced within a user's reference template being a significant inhibiting factor. It is suggested, however, that sufficient discriminative traits exist within our day-to-day interactions to authenticate a person (Fig. 4.26).

Although not unique and distinct enough to be used within an identification system, this technique is non-intrusive and can be used to continuously monitor the identity of users whilst they work on their computer system. However, this very advantage also has a disadvantage with regard to users' privacy, as their actions will be continually monitored, and such information has the potential to be misused (e.g. the capturing of passwords). Acceptability is therefore going to be determined by how these two factors play out in practice. Although no authentication mechanisms exist utilising this technique, a number of companies are utilising profiling as a means of fraud protection on credit card and mobile telephony systems (Gosset 1998; Stolfo et al. 2000). Within these systems, which are very specific to a subset of tasks, studies have reported detection rates exceeding 90% with false alarm rates as low as 3% (Stormann 1997).

4.4.4.2 Gait Recognition

This process utilises the way in which an individual walks to determine their identity. Gait recognition has an obvious advantage from the collection perspective in that it can be achieved from quite a distance, more so than any other biometric approach (most of which require physical contact with the sensor). This leads to it being used almost exclusively for identification purposes (as no opportunity to provide a claimed identity is available) and several unique applications where discrete monitoring of the individuals is required, such as airports. However, like all behavioural biometrics, external factors play a significant role in the variance of samples. Footwear, walking surface and clothing all play a role, as well as indoor versus outdoor and the mindset of the person. Gait recognition also suffers from issues with regard to its universality and permanence. It would not work on those who are wheelchair-bound and disabled. Ignoring the already fairly large day-to-day variance that exists in our gait due to carrying bags and environmental facts, long-term recognition will also be an issue as age and illness play a role.

The process of classification is generally categorised into two approaches: shape and dynamics. A gait sample includes the walking of an individual over two strides. The shape-based approach looks at the overall image shape of the individual over this cycle, whereas the dynamics approach looks at the rate of transition within the cycle. Performance rates vary considerably depending upon the study but identification rates can range from 78% to 3% depending upon the dataset (Sarkar et al. 2005).

4.4.4.3 Keystroke Analysis

The way in which a person types on a keyboard has been shown to demonstrate some unique properties (Spillane 1975). The process of authenticating a person from their typing characteristic is known as keystroke analysis (or dynamics). Authentication itself can be performed in both *static* (text-dependent) and *dynamic* (text-independent) modes, the former being the more reliable approach. The particular characteristics used to differentiate between people can vary, but often include the time between successive keystrokes, also known as the inter-keystroke latency and the hold time of a key press.

The unique factors of keystroke analysis are not discriminative enough for use within an identification system, but can be used within a verification system.

However, although much research has been undertaken in this field due to its potential use in computer security, only one commercial product to date has been deployed and is based on the former (and simpler) method of static verification. Sentry (formally Biopassword) performs authentication based upon a person's username and password (Scout Analytics 2011). A major downside to keystroke analysis is the time and effort required to generate the reference template. As a person's typing characteristics are more variable than say a fingerprint, the number of samples required to create the template is greater, requiring the user to repetitively enter a username and password until a satisfactory quality level is obtained. The performance of keystroke analysis varies considerably amongst studies, but a notable study by Joyce and Gupta (1990) managed to achieve an FRR of 16.36% with an FAR of 0.25% using a short string.

4.4.4.4 Signature Recognition

As the name implies, signature recognition systems attempt to authenticate a person based upon their signature. Although signatures have been used for decades as a means of verifying the identity of a person on paper, their use as a biometric is more recent. The use of touch-screen interfaces such as those on PDAs, mobile phones and tablets has made the acquisition of samples simpler (and cheaper). Authentication of the signature can be performed statically and/or dynamically. Static authentication involves utilising the actual features of a signature, whereas dynamic authentication also uses information regarding how the signature was produced, such as the speed and pressure. The latter approaches are far more robust against forgery – an attack that has always plagued signature-based verification. Commercial applications exist, including for use within computer access and point-of-sale verification, and are frequently utilised in the USA. Performance rates for signature recognition are better than most behavioural approaches with an EER of 2.84% being reported (Yeung et al. 2004).

4.4.4.5 Speaker Recognition (or Voice Verification)

A natural biometric, and arguably the strongest behavioural option, voice verification utilises many physical aspects of the mouth, nose and throat, but is considered a behavioural biometric as the pronunciation and manner of speech is inherently behavioural. Although similar, it is important not to confuse voice verification with voice recognition, as both systems perform a distinctly different task. Voice recognition is the process of recognising *what* a person says, whereas voice verification is recognising *who* is saying it. Voice verification, similar to keystroke analysis, can be performed in static (text-dependent) and dynamic (text-independent) modes, again with the former being a simpler task than the latter. Pseudo-dynamic approaches do exist which request the user, who has not been explicitly trained during enrolment, to say two numbers randomly. Numerous companies exist

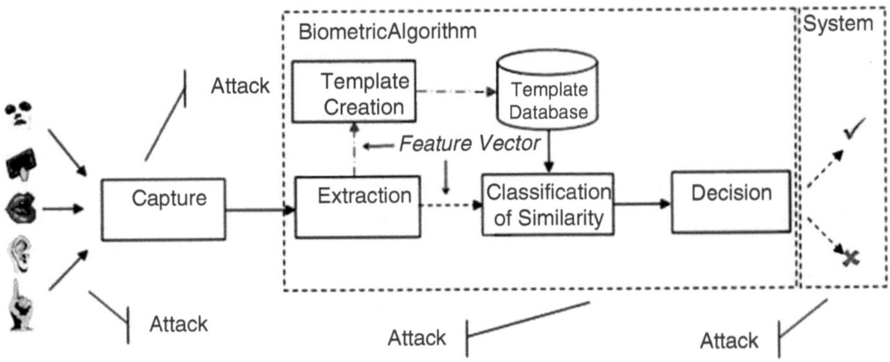

Fig. 4.27 Attacks on a biometric system

providing various applications and systems that utilise voice verification, for instance, Nuance (2011) provides static authentication for call centres. In terms of performance, speaker recognition is typically amongst the best-performing behavioural approach with an EER of approximately 2% (Przybocki et al. 2007). However, the authors from this study also demonstrate the viability of the performance characteristics depending upon the nature of the sample.

The list of biometrics provided should not be considered exhaustive as new techniques and measurable characteristics are constantly being identified. The common underlying trend that appears within behavioural approaches is that their performance is highly variable depending upon the scenario and external factors. The levels of uniqueness and permanence in comparison to their physical counterparts are poorer. However, if authentication is no longer about providing a point-of-entry decision with an apparent Boolean decision, behavioural approaches have the advantage of generally being more usable, acceptable and less intrusive. Either way, care is required in the design and implementation of the approach in order to manage these various issues effectively.

4.4.5 Attacks Against Biometrics

As with all authentication approaches, efforts have been made to hack, break and circumvent biometric systems. As these systems tend to be more complex than their secret-knowledge and token counterparts, in that they contain more components and rely upon probabilistic outcomes, they do offer a wider scope for possible misuse.

The simplest way to categorise the possible attack vectors against a biometric is to consider all the components that make up the system. In reality, each component – the sample, the sensor, the biometric algorithm, the system or any of the communication links in between – can possibly be compromised. Figure 4.27 illustrates the various attack vectors against a biometric system.

Fig. 4.28 USB memory with fingerprint authentication

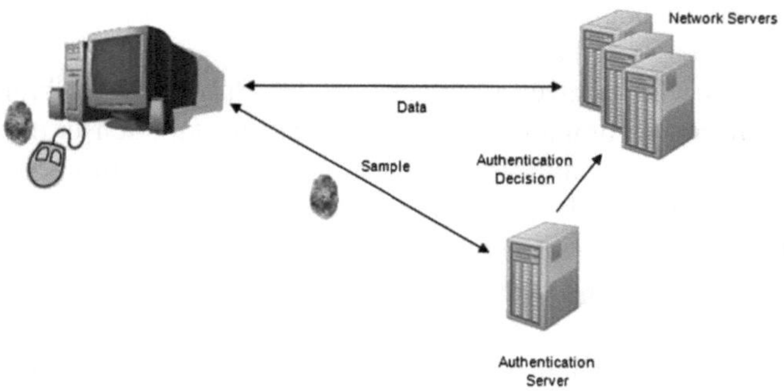

Fig. 4.29 Distributed biometric system

The ability and difficulty for an attacker to compromise an element will greatly depend upon the implementation of the individual biometric. Biometric systems can come in a variety of forms from completely standalone devices that perform the capture, extraction, classification and decision process internally or locally (e.g. a USB memory device with a fingerprint sensor – as illustrated in Fig. 4.28) to systems that perform various aspects of the process in physically disparate systems. For instance, an enterprise deployment of fingerprints to replace standard logins could have distributed functionality throughout the network. An example is illustrated in Fig. 4.29. The sensor on the mouse or keyboard captures the biometric sample. The sample is then communicated to an authentication server via the desktop system.

The authentication server will perform extraction and classification – possibly on multiple servers to allow for load balancing which is required when large volumes of authentication are necessary (e.g. when everyone arrives at 9 a.m. to start work). The decision from the authentication server is then communicated to the enterprise server that controls access to all systems. That decision is then filtered down to the desktop system, which will subsequently log in or deny access. The latter system has far more scope for misuse simply due to the larger number of systems that are involved and the various communication links between the systems that provide an opportunity for interception.

Standalone systems are typically far more resistant to attack as the complete process has been deployed into a single processing unit. An opportunity to sniff or capture information from bus lines (metal tracks and wires connecting components) or read the memory off chips is more limited. Whilst they have good tamper resistance, poor design still offers opportunities for misuse, with a number of USB memory keys with fingerprint sensors having been hacked.

Considering each of the components in turn, the acquisition of a biometric sample via forceful means is an inevitable consequence, largely due to its simplicity versus the remaining more technical attacks. Fortunately, to date, few such attacks have occurred; however, one notable incident in Malaysia did result in a person losing his thumb (Kent 2005). The communication links are also an obvious target for attack. However, approaches to attacking and defending against these forms of attack fall under the standard domain of network security (i.e. end-to-end cryptographic support) and therefore will not be considered.

Instead focus will be given to attacks on the sensor and the biometric algorithm. By far the most common attacks are those against the sensor itself. Spoofing is a technique where the attacker is able to provide a fake biometric sample to the sensor and convince it that it is indeed a legitimate sample. Attacking the system at the presentation stage removes many of the technical barriers that exist when compromising other aspects of the system. The nature of some biometric approaches is that either fragments or complete copies are left behind. For instance, fingerprints are left behind on many objects people touch. An attacker can use the same techniques deployed by law enforcement agencies for lifting the prints. Once lifted, a duplicate can be created from freely available modelling materials (Matsumoto et al. 2002). Figure 4.30 illustrates two examples of fake fingers created using silicon and jelly. The second image illustrates a successful capture and authentication using a fake silicon finger. Facial images are created every time a photograph is taken of you. A simple printout can be used to spoof a sensor, as illustrated in Fig. 4.31. There have also been documented cases of attackers simply breathing on a fingerprint sensor and obtaining access – this works due to the latent oils from the finger remaining on the sensor from the previous person (Dimitriadis and Polemi 2004).

Behavioural biometrics techniques are subject to forgery, particularly if an action is observed previously. Signature recognition, keystroke analysis, voice verification and gait recognition all have the opportunity of being misused in this way. Although, as with traditional forgery, it often requires a specialist skill in its own right in order

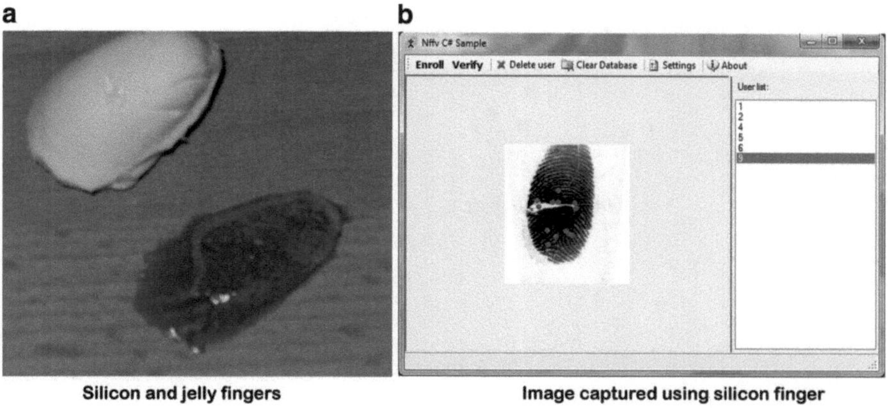

Fig. 4.30 Examples of fake fingerprint

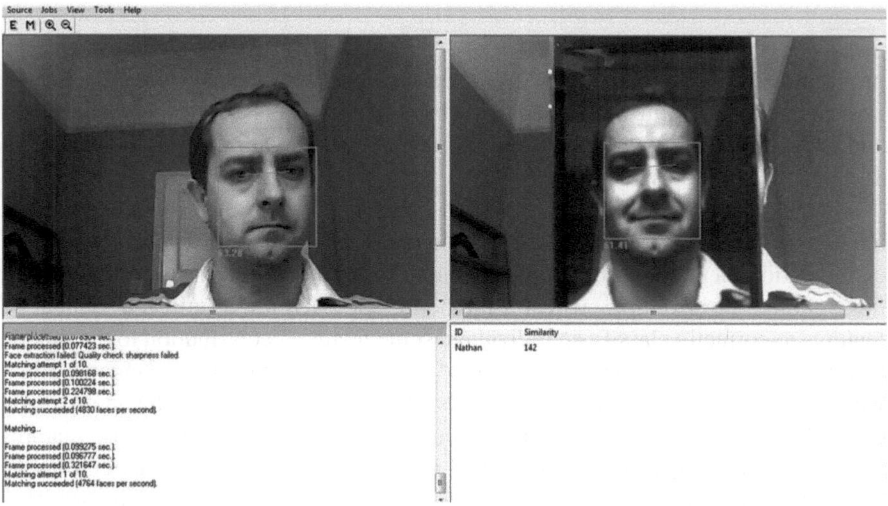

Fig. 4.31 Spoofing facial recognition using a photograph

to mimic the behaviour of the authorised individual, for example, the way a TV impersonator is able to mimic the voice and mannerisms of celebrities.

As a consequence, biometric vendors are adding a level of sophistication to the approaches that assist in minimising the threat posed by spoofing. For many approaches, a liveness component is included in the system to ensure the sample being provided is acquired from a live host. Approaches taken for fingerprint systems include temperature, pulse, blood pressure, perspiration and electrical resistance. Facial recognition systems have utilised approaches such as rapid eye movement and 3D imagery to determine liveness. Whilst practically, different biometric techniques have differing levels of liveness implemented, some have fallen short in

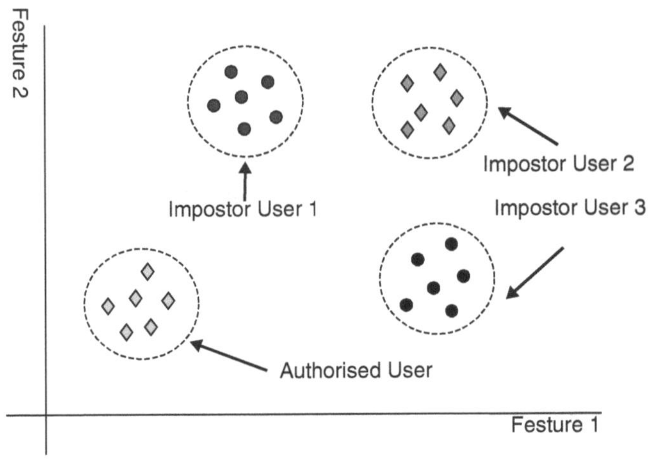

Fig. 4.32 Diagrammatic demonstration of feature space

providing effective protection, with ingenious attackers circumventing even these measures. However, this continued battle between designers and hackers will persist and liveness detection will improve in time and become an inherent component of all biometric products.

Attacks against the biometric algorithm can be amongst the most technically challenging to achieve. A thorough analysis and understanding of the data extraction and classification elements can highlight possible weaknesses in the algorithm that can be exploited. For example, whilst a biometric approach might utilise a number of features to correctly classify an individual, some features' contribution to the unique discriminatory information that is provided could be more than others. Understanding which play a more important role provides a focus for the attacker. This information could then be either used to produce a sample that would conform to those features or more likely be injected into the system as a mechanism for bypassing the sensor.

Another approach to this type of attack is to analyse the performance characteristics of an approach at the individual user level. Whilst performance metrics are given in various forms such as EER, FAR or FRR, in reality for statistical reliability, these figures are actually average performances across a population of people. An examination of individual performance characteristics (the match score) will often highlight impostors with very high FAR against a particular authorised user. A demonstration of this is illustrated in Fig. 4.32. These users have characteristics that are more similar to the authorised user than others. The collection and analysis of these will highlight potential weaknesses in the system.

These more technically challenging attacks do assume the attacker has free access to the technology – either the product or a software development kit (SDK) of the algorithm, so that repeated samples can be sent through the process and analysed. The final, rather non-technical approach is to simply attempt a brute-force attack against the feature vector. After all, the feature vector is simply a form of secret knowledge. The key difference with biometric samples is that the attacker does not need to reproduce the feature vector exactly, but just with sufficient similarity for the system to deem it an authorised sample. How similar depends upon the configuration of the threshold value and the level of security the administration wants to achieve.

4.5 Summary

Authentication is key to maintaining security and a wide range of approaches have been developed to meet the unique needs of systems that have been created. At its purest form, authentication is all about a secret and comparing a sample against the secret. Passwords are therefore theoretically the strongest form of authentication and can provide more than sufficient protection – assuming of course they don't rely upon people to remember them. An almost natural evolution has occurred with the identification of weaknesses in passwords, development of tokens to take the cognitive pressure off people, recognition that tokens only authenticate the presence of the token – rather than the person in possession of it – and finally biometrics, an authentication credential linked to a person, removing any reliance upon them having to remember, but merely utilising a natural consequence of our being.

Biometrics, however, are not the panacea of the authentication problem. They introduce their own set of issues, whether it be weaknesses in individual approaches, the lack of universality resulting in administrators having to deploy and manage multiple systems, poor permanence, low acceptability, the need for complex systems to collect and process samples or concerns related to privacy. However, biometrics do offer an ability to disconnect the cognitive load from the user and secure it within a container that provides far more effective security than a token is able to achieve. But that in itself will not be sufficient.

Due to the aforementioned weaknesses discussed throughout this chapter, of all authentication approaches, correct authentication of the user can never be guaranteed. Instead, organisations have to live with the risk that compromise can occur and they deploy additional countermeasures to mitigate that risk. Unfortunately, those risks are not always truly understood, and pressure is mounting on the need for more reliable security countermeasures. Foremost to this need are effective, reliable and usable authentication technologies. For this to happen, the concept of authentication as a point-of-entry approach must be re-evaluated.

References

Ashbourn, J.: Biometrics: Advanced Identity Verification: The Complete Guide. Springer, London (2000). ISBN 978-1852332433

Balaban, D.: Transport for London to Discard Mifare Classic. NFC Times. Available at: http://www.nfctimes.com/news/transport-london-discard-mifare-classic-seeks-desfire-sims (2010). Accessed 10 Apr 2011

Bank of America: SiteKey at Bank of America. Bank of America. Available at: http://www.bankofamerica.com/privacy/index.cfm?template=sitekey (2011). Accessed 10 Apr 2011

BBC: Personal data privacy at risk. BBC News. Available at: cbra http://news.bbc.co.uk/1/hi/business/7256440.stm (2008). Accessed 10 Apr 2011

Blonder, G.E.: Graphical passwords. U.S. Patent 5559961, Lucent Technologies Inc, Murray Hill, 1995

Brostoff, S., Sasse, M.A.: Are Passfaces more usable than passwords? A field trial investigation. In: Proceedings of Human Computer Interaction, Sunderland, pp. 405–424 (2000)

Chip and PIN: Why did we change. Chip and PIN. Available at: http://www.chipandpin.co.uk/consumer/means/whychanging.html (2006). Accessed 10 Apr 2011

Crown Copyright: Using the Iris recognition immigration system (IRIS). Crown Copyright. Available at: http://www.ukba.homeoffice.gov.uk/travellingtotheuk/Enteringtheuk/usingiris/ (2010). Accessed 10 Apr 2011

Das, R.: Retina recognition: biometric technology in practice. Keesing Journal of Documents and Identity, issue 22. Available at: http://www.biometricnews.net/Publications/Biometrics_Article_Retinal_Recognition.pdf (2007). Accessed 10 Apr 2011

Daugman, J.: Biometric personal identification system based on Iris Recognition. US Patent 5,291,560 (1994)

de Winter, B.: New hack trashes London's Oyster card. Tech World. Available at http://news.techworld.com/security/105337/new-hack-trashes-londons-oyster-card/ (2008). Accessed 10 Apr 2011

Dimitriadis, C., Polemi, D.: Biometric authentication. In: Proceedings of the First International Conference on Biometric Authentication (ICBA). Springer LNCS-3072, Berlin/Heidelberg (2004)

FVC2006: Open category: average results over all databases. Biometric System Laboratory. Available at: http://bias.csr.unibo.it/fvc2006/results/Open_resultsAvg.asp (2006). Accessed 10 Apr 2011

Gosset, P. (eds.): ASPeCT: Fraud detection concepts: final report. Doc Ref. AC095/VOD/W22/DS/P/18/1 (1998 Jan)

Hurley, D., Nixon, M., Carter, J.: Force field feature extraction for ear biometrics. Comput. Vis. Image Understand. **98**, 491–512 (2005)

IBG.: How is biometrics defined? International Biometrics Group. Available at: http://www.biometricgroup.com/reports/public/reports/biometric_definition.html (2010a). Accessed 10 Apr 2011

Ingersoll-Rand.: HandKey. IR Security Technologies. Available at: http://w3.securitytechnologies.com/Products/biometrics/access_control/handkey/Pages/default.aspx (2011). Accessed 10 Apr 2011

IrsisGuard.: Iridian Announces UAE border control system exceeds one million transactions IrisGuard. Available at: http://www.irisguard.com/pages.php?menu_id=&local_type=5&local_id=1&local_details=1&local_details1=&localsite_branchname=IrisGuard (2004). Accessed 10 Apr 2011

ISO.: ISO/IEC 7813:2006 Information Technology – Identification Cards – Financial Transaction Cards. International Standards Organisation. Available at http://www.iso.org/iso/iso_catalogue/catalogue_tc/catalogue_detail.htm?csnumber=43317 (2006). Accessed 10 Apr 2011

ISO: ISO JTC 1/SC37 – Biometrics. International Standards Organisation. Available at: http://www.iso.org/iso/iso_catalogue/catalogue_tc/catalogue_tc_browse.htm?commid=313770&published=on&development=on (2010). Accessed 10 Apr 2011

Jain, A., Patrick, F., Arun, R.: Handbook of Biometrics. Springer, New York (2008). ISBN 978-0-387-71040-2

Joyce, R., Gupta, G.: Identity authentication based on keystroke latencies. Commun. ACM **39**, 168–176 (1990)

Kent, J.: Malaysia car thieves steal finger. BBC News. Available at: http://news.bbc.co.uk/1/hi/world/asia-pacific/4396831.stm (2005). Accessed 10 Apr 2011

Maltoni, D., Maio, D., Jain, A., Prabhakar, S.: Handbook of Fingerprint Recognition. Springer, New York (2005). ISBN 978-0387954318

Matsumoto, T., Matsumoto, H., Yamada, K., Hoshino, S.: Impact of artificial 'gummy' fingers on fingerprint systems. Proc. SPiE **4677**, 275–289 (2002)

Miura, N., Nagasaka, A., Miyatake, T.: Feature extraction of finger-vein patterns based repeated line tracking and its applications to personal identification. Mach. Vis. Appl. **15**, 194–203 (2004)

Moreno, B., Sanchez, A.: On the use of outer ear images for personal identification in security applications. In: Proceedings of IEEE 33rd Annual International Conference on Security Technologies, Madrid, pp. 469–476 (1999)

Nanavati, S., Thieme, M., Nanavati, R.: Biometrics Identity Verification in a Networked World. Wiley, New York (2002). ISBN 0471099457

NSTC: Biometrics glossary. National Science and Technology Council. Available at: http://www.biometrics.gov/Documents/Glossary.pdf (2006). Accessed 10 Apr 2011

Oechslin, P.: Making a faster cryptoanalytic time-memory trade-off. In: Advances in cryptology – CRYPTO 2003, 23 rd Annual International Cryptology Conference, Santa Barbara August 17–21, 2003, Proceedings. Lecture Notes in Computer Science 2729. Springer 2003, Berlin/Heidelberg, ISBN 3-540-40674-3 (2003)

Oxford University Press: How many words are there in the English language. Oxford University Press. Available at http://www.oxforddictionaries.com/page/93 (2010). Accessed 10 Apr 2011

Paivio, A., Rogers, T.B., Smythe, P.C.: Why are pictures easier to recall than words? Psychon. Sci. **11**, 137–138 (1968)

Passfaces: Passfaces: two factor authentication for the enterprise. Passfaces Corporation. Available at http://www.realuser.com/index.htm (2011). Accessed 10 Apr 2011

Phillips, J., Scruggs, T., O'Toole, A., Flynn, P., Bowyer, W., Schott, C., Sharpe, M.: FRVT 2006 and ICE 2006 large-scale results. NIST IR 2007. Available at: http://face.nist.gov/frvt/frvt2006/FRVT2006andICE2006LargeScaleReport.pdf 2007. Accessed 10 Apr 2011

Przybocki, M., Martin, A., Le, A.: NIST speaker recognition evaluations utilising the mixer corpora – 2004, 2005, 2006. IEEE Trans. Audio Speech Lang. Process **15**(7), 1951–1959 (2007)

RSA: Securing your future with two-factor authentication. EMC Corporation. Available at: http://www.rsa.com/node.aspx?id=1156 (2011). Accessed 10 Apr 2011

Sarkar, S., Phillips, P., Liu, ., Robledo-Vega, I., Grother P, Bowyer, K.: The Human ID gait challenge problem: data sets, performance and analysis. IEEE Trans. Pattern. Anal. Mach. Intell. **II**, 162–177 (2005)

Scout Analytics: Sentry: zero footprint, strong authentication. Scout Analytics. Available at: http://www.biopassword.com/zero_footprint_strong_authentication.asp (2011). Accessed 10 Apr 2011

Shepard, R.N.: Recognition memory for words, sentences, and pictures. J. Verbal Learn Verbal Behav **6**, 156–163 (1967)

Smith, R.: Authentication: From Passwords to Public Keys. Addison and Wesley, Boston (2002). ISBN 0201615991

Socolinsky, D., Selinger, A.: Face detection with visible and thermal infrared imagery. Comput. Vis Image Understand, pp. 72–114, July–August (2003)

Socolinsky, D., Selinger, A.: Thermal face recognition in an operational scenario. In: CVP04, Washington, DC, pp. 1012–1019 (2004)

Spillane, R.: Keyboard apparatus for personal identification. IBM Tech. Disclosure Bull. **17**, 3346 (1975)

Stolfo, S.J., Wei F., Wenke L., Prodromidis, A., Chan, P.K.: Cost-based modeling for fraud and intrusion detection: results from the JAM project. In: DARPA Information Survivability Conference and Exposition, 2000. DISCEX '00. Proceedings, vol. 2, Hilton Head, pp. 130–144 (2000)

Stormann, C.: . Fraud management tool: evaluation report. Advanced Security for Personal Communications (ASePECT), Deliverable. 13, Doc Ref. AC095/SAG/W22/DS/P/13/2 (1997)

Symantec: SennaSpy generator. Symantec Corporation. Available at http://www.symantec.com/security_response/writeup.jsp?docid=2001-062211-2540-99 (2007). Accessed 10 Apr 2011

Thorpe, J., van Oorschot, P.C.: Human-seeded attacks and exploiting hot-spots in graphical passwords. In: 16th USENIX Security Symposium, Boston, pp. 103–118 (2007)
Vuagnoux, M., Pasini, S.: Compromising electromagnetic emanations of wired and wireless keyboards. In: 18th USENIX Security Symposium, Montreal, pp. 1–16 (2009)
Wikimedia Commons: Welcome to Wikimedia, Wikimedia Commons. Available at: http://commons.wikimedia.org/wiki/Main_Page (2011). Accessed 10 Apr 2011
Wireshark: Wireshark. Wireshark Foundation. Available at http://www.wireshark.org/ (2011). Accessed 10 Apr 2011
Woodward, J., Orlans, N., Higgins, P.: Biometrics and Strong Authentication. McGraw-Hill, Berkeley (2003). ISBN 978–0072222272
Yeung, D., Chang, H., Xiong, Y., George, S., Kashi, R., Matsumoto, T., Rigoll, G.: SVC2004: first international signature verification competition. In: Proceedings of ICBA. Springer LNCS-3072, Berlin/Heidelberg, pp. 16–22 (2004)

Chapter 5
Transparent Techniques

5.1 Introduction

The nature of authentication currently is to authenticate at point-of-entry. Therefore all authentication systems have been designed and developed to tightly operate within those requirements. Applying those same technologies to a different application, such as transparent authentication, results in a series of challenges. This is not completely unexpected, as systems designed for one application are invariably not fit-for-purpose within another application, without some redesign. For biometric-based approaches, the primary challenge is concerned with the sample capture. Within point-of-entry scenarios, the environment within which the sample is taken can be closely controlled whether it be an application guiding the user to provide the sample, swiping the finger at a slower pace across the sensor, changing the orientation of the head so that the camera has a better shot of the face or environment-based controls, such as minimising ambient noise or ensuring an acceptable level of illumination. Biometric systems have therefore been developed knowing such operational variables are under their control – or remain constant. With transparent authentication, such control over the environment and the user is not possible. Instead, the user could be performing a variety of activities whilst the sample is captured, giving rise to a far higher degree of variability in the samples.

A caveat exists in that some biometric systems are designed for non-intrusive applications, but these tend to be covert applications performed at a distance. Watchlists are utilised not to authenticate a user but rather to identify an individual and to do so from a limited subset of the population (e.g. criminals or terrorists). For instance, feeds from CCTV cameras can be used to perform facial recognition. Whilst some of the issues surrounding successful watchlist applications are also applicable to transparent authentication, such as facial orientation, that is where the comparison ends. The majority of transparent-enabled approaches are not suited to watchlist applications, effectively a form of identification rather than verification system.

Transparency is not simply connected to biometrics techniques, but can indeed includes other forms of authentication. Whilst it is difficult to envisage how secret

N. Clarke, *Transparent User Authentication: Biometrics, RFID and Behavioural Profiling*, DOI 10.1007/978-0-85729-805-8_5, © Springer-Verlag London Limited 2011

knowledge–based approaches could be used transparently, some token-based approaches exist that would be feasible, such as radio frequency identification (RFID)-based approaches. For example, an item of personal jewellery worn on the same hand as the mobile device could provide a basis for ensuring only the legitimate user is given access.

The application of transparent and continuous authentication is not universal across all scenarios. Indeed, where one-off authentication is necessary, point-of-entry authentication continues to be a viable option. Transparent and continuous authentication is useful in application scenarios that require persistent verification of the user. Logical access to computing systems, whether desktop PCs, laptops or mobile phones, would be a suitable scenario.

The following sections will closely examine a number of techniques that lend themselves to transparency. Whilst not a conclusive list, the accompanying discussion also highlights the issues that exist with point-of-entry approaches to authentication and suggests techniques to enable transparency. In particular, research studies carried out by the author and his colleagues in the Centre for Security, Communications and Network Research (CSCAN) at Plymouth University will be utilised to illustrate how transparency can be applied.

5.2 Facial Recognition

Facial recognition has become an increasingly popular approach to utilise in a wide variety of applications. It is one of the few physiological approaches that are applicable for being directly applied in a transparent fashion and the reason for this is that the sensor is able to capture a sample without the user specifically needing to explicitly provide a sample. In many other cases such as fingerprint recognition, hand geometry and vascular pattern recognition, the user must explicitly touch or place their hand on a sensor.

The use of facial recognition to date has typically focused upon very well-defined environments, with controls or restrictions placed upon the illumination, facial orientation and distance from the capture device. Of course, the ability for the sensor, a camera in this case, to provide a useful sample is different to merely providing an image. With no user interface or instruction, the sensor is required to capture the image of a face with sufficient clarity that a facial recognition algorithm can be utilised. Depending upon the scenario, this is a more challenging proposition:

- For desktop environments, as the PC will physically reside in a same location, the environment and external factors, such as distance from camera and facial orientation from camera, can all be controlled to a fairly good degree.
- For laptop environments, the user could be in a variety of physical locations when using his computer: home, work or travelling. Therefore environmental factors such as illumination are going to vary. External factors, such as distance from the screen and facial orientation, should remain fairly constant if the user is

Fig. 5.1 Environmental and external factors affecting facial recognition

typing – in order for the user to type effectively, the hands and screen need to be positioned in a particular fashion, thereby providing consistent spacing.

- For mobile device environments, both the environment and external factors are likely to be far more variable. Rather than constraining the environment to perhaps a small subset of locations such as home, work or airport, a mobile device will be used in a wide variety of environmental situations: inside, outside, whilst walking, during the day and at night. The external factors are also likely to be more variable as the user could be performing a variety of operations on the device, using one hand or two, standing up or sitting down. All will vary the distance of the handset to the face and provide differing facial orientations.

These environmental and external factors are important for the success of facial recognition. Correct illumination is required in order for extraction algorithms to be able to identify the correct facial features for measurement. Distance from the camera and facial orientation will cause (if not accounted for) the measurements to vary. As illustrated in Fig. 5.1, the ability to determine the correct measurements can prove challenging.

In order to address the issue of transparency, it would be possible to simply implement facial recognition as it is and enable the camera to capture samples whilst the user is interacting with the device – as this will provide a mechanism for ensuring someone is in front of the camera. An image quality measure could then be used during the feature extraction phase to determine whether the sample is of sufficient quality to be used for authentication – a process that is present in many biometric systems currently. However, an image quality measure alone will not suffice, as it will either simply respond with failures on the image quality, or with false decisions at the matching stage as the feature vector will not be appropriately similar.

To improve the tolerance of the technique to variations in environment and external factors, two options are available:

- First, to undertake research looking into improving the extraction and classification algorithms with a view to removing the dependence upon these factors
- Second, to look to adapt current classification algorithms in a fashion that achieves transparency

Whilst the former approach would be a sensible one it requires significant investment of resources by biometric vendors. The latter approach simply proposes utilising the existing approach but in a different manner.

Developments in improving facial recognition with respect to environmental and external factors are taking place. This is not simply a problem for transparent application (although it is exacerbated) but also for traditional facial recognition applications. The most significant effort being made to date is in the area of three-dimensional (3D) facial recognition. With two-dimensional (2D) facial images, differing illuminations cause differing shadows to fall on the face, which obscures the facial characteristics and subsequent measurements. Three-dimensional approaches develop a full mapping of the facial contours, thereby providing a more robust and consistent set of measurements invariant to changes in pose or illumination. Compared to traditional 2D facial recognition, 3D is far newer; however, initial studies have proven relatively successful with an 86% rank 1 correct recognition rate (Lu et al. 2006). Unfortunately, one of the key aspects of enabling transparent authentication is the use of sensor technology that could reasonably be expected to be present. Whilst web cameras are fairly ubiquitous across computing devices, 3D cameras have yet to be deployed in the mainstream, largely due to prohibitively high costs. The sensors are also rather bulky and unwieldy for use in discrete capture. This approach could therefore certainly be appropriate for transparent authentication but only with future developments in sensor technology.

Research has also been undertaken looking into how effective existing algorithms can be when applied to transparency. Research carried out by the author has shown promise in manipulating current algorithms (Clarke et al. 2008). The commentary that follows provides a brief overview of the approach.

The proposed method of adapting existing algorithms is to move away from a one-to-one comparison of an image with a template (as depicted in Fig. 5.2), and replace the template with a series of images that represent various facial orientations of the authorised user (as illustrated in Fig. 5.3). In this way, existing pattern classification algorithms can still be applied, but the overall approach should be more resilient to changes in facial orientation. Indeed, traditional facial recognition mechanisms sometimes use this approach, whereby they collect a series of front-facing samples and verify against each in turn, rather than depending upon

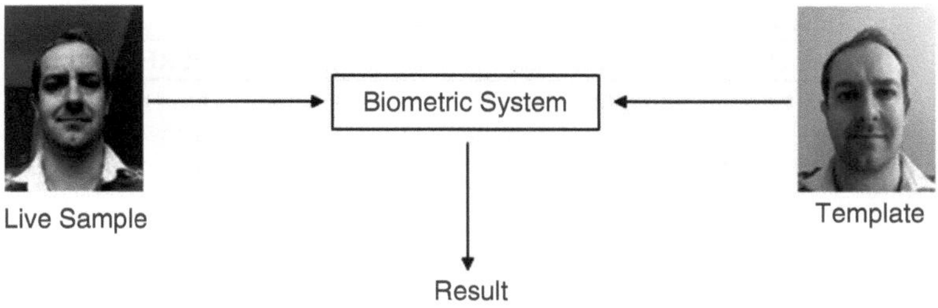

Fig. 5.2 Normal facial recognition process

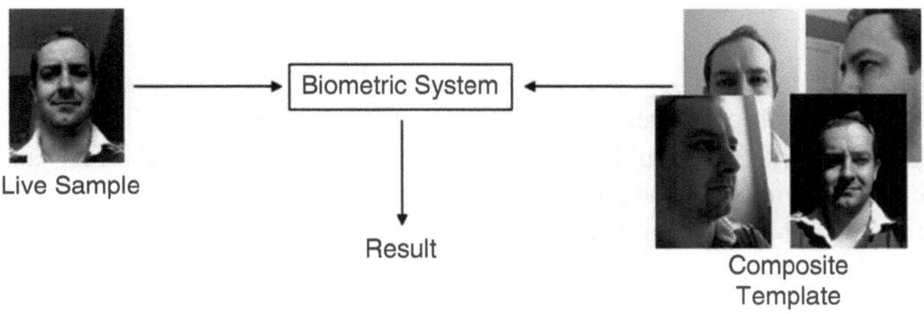

Fig. 5.3 Proposed facial recognition process

a single sample. However, they have not been applied to such large variations in facial orientation previously.

Under this proposed mechanism, each sample will effectively be compared to a series of images stored within a composite template and the number of verifications will subsequently increase. This will therefore introduce an increased likelihood that an impostor is accepted by an appropriate similarity with at least one of the series of images. For each additional verification performed, the system is providing a further opportunity for an impostor to be accepted. From a performance perspective, under this proposed system, the false acceptance rate (FAR) will only ever be as good as the original FAR of the algorithm being used (under the normal facial recognition process), with more realistically an increase in the FAR being experienced. Conversely, however, under this proposed system the false rejection rate (FRR) will at worst equal that of the previous FRR, but more realistically be lower because additional poses are present within the composite template.

Effectively, this process is looking to trade off one performance metric for the other, reducing the level of security for an improvement in the level of usability. The advantage of trading off the FAR and FRR in facial recognition is twofold:

1. Facial recognition approaches have quite distinctive characteristics and experience good levels of performance in terms of FAR and FRR. Indeed, facial recognition systems are often used in identification systems as well as verification systems. Their use for verification does not require such a high degree of distinctiveness and can arguably be traded off for usability.

Table 5.1 Subset of the FERET dataset utilised

Dataset Ref	Description	Angle	FERET Ref
1	Front face	0	ba
2	Alternative front face	0	bj
3	Left image	+60	bb
4	Left image	+40	bc
5	Left image	+25	bd
6	Left image	+15	be
7	Right image	−15	bf
8	Right image	−25	bg
9	Right image	−40	bh
10	Right image	−60	bi

2. The relationship between the FAR and FRR is non-linear. Therefore small changes in the FAR could possibly result in larger changes in the FRR and conversely. Obviously if the converse were to occur, this approach is less likely to be implementable – it would all depend on the level of security the algorithm could provide for the level of usability.

An experiment was therefore conducted looking to establish whether it is possible to take advantage of these properties to provide a little less security for a larger improvement in the robustness and usability of the approach. Three aspects were considered:

1. A control experiment where the facial recognition system would be tested under normal conditions
2. An experiment to evaluate the effect upon the performance rates when using images of varying facial orientation against a normal template
3. An experiment to evaluate the effect upon the performance rates when using images of varying facial orientation against the proposed composite template

Whilst a wide variety of facial datasets exist, including the FERET colour dataset, the YALE dataset, PIE dataset, AT&T dataset, MIT dataset and NIST Mugshot Identification dataset to name but a few (Gross 2005), the FERET colour dataset was utilised as this contained the varying facial orientations required for this set of experiments (as shown in Table 5.1).

The methodology employed for the study followed the standard research methodology for biometric experiments in that each of the participants took a turn playing the role of the authorised user, with all remaining users acting as impostors. The images used for creating the biometric template, whether it is the normal or composite template, are not used in the evaluation phase. Specific details of the datasets used in each experiment are given in Table 5.2.

It was anticipated that some facial recognition algorithms would be more tolerable of changes in facial orientations than others. Therefore a series of algorithms were selected that represent a number of established facial recognition algorithms. The algorithms themselves were obtained from Advanced Source Code (Rosa 2008).

Table 5.2 Datasets utilised in each experiment

	Enrolment		Verification	
Exp	Dataset ref	No. of participants	Dataset ref	No. of participants
1	1	200	2	200
2	1	200	2, 3, 4, 5, 6, 7, 8, 9, 10	200
3	1, 3, 5, 8, 10	200	2, 4, 6, 7, 9	200
4	1, 3, 5, 8, 10	50	2, 4, 6, 7, 9	150

Table 5.3 Facial recognition performance under normal conditions

Algorithm	FRR (%)	FAR (%)
Fourier-Bessel	31.5	0.16
Fisherfaces	21	0.11
Fourier-spectra	24.5	0.12
Gabor filters	4.5	0.023

Table 5.4 Facial recognition performance with facial orientations

Algorithm	FRR (%)	FAR (%)
Fourier-Bessel	50.8	0.25
Fisherfaces	31.8	0.16
Fourier-spectra	38.3	0.19
Gabor filters	46.2	0.23

The underlying performance of the five algorithms, as outlined by experiment 1 and shown in Table 5.3, suggests each of the algorithms are effective in rejecting impostors with a FAR of 0.19% or below. Unfortunately, the accompanying FRR is considerably larger with error rates between 21% and 39% for four of the techniques. The only technique that performed well was Gabor Filters with a FRR of 4.5%. It should be noted that the larger values of FRR could be a result of fewer actual verification samples as compared to the impostors. For instance, for each user, the FRR is based upon a single image, resulting in an FRR of either 0% or 100%, whereas the FAR is based upon 199 other images, resulting in the FAR increasing in steps of 0.5%. This is unfortunately a result of the lack of repeated images per user for each orientation in the dataset. Nevertheless, the purpose of this experiment was to largely understand and establish the level of security rather than usability of the underlying classification algorithms.

When applying the additional facial orientations to the verification process, increasing the number of images utilised to calculate the FRR from 1 to 5 (and thereby mitigating some of the effect that a lack of comparisons can cause) the FRR increases across all algorithms. Even the Gabor–Filters approach that achieved a 4.5% FRR in experiment 1 has now increased to 46.2%. The FAR has marginally increased across the five algorithms, but is still able to provide a good level of security against impostors (Table 5.4).

The results from this experiment demonstrate the inability of current facial recognition algorithms to cope with input samples that have a high degree of variability

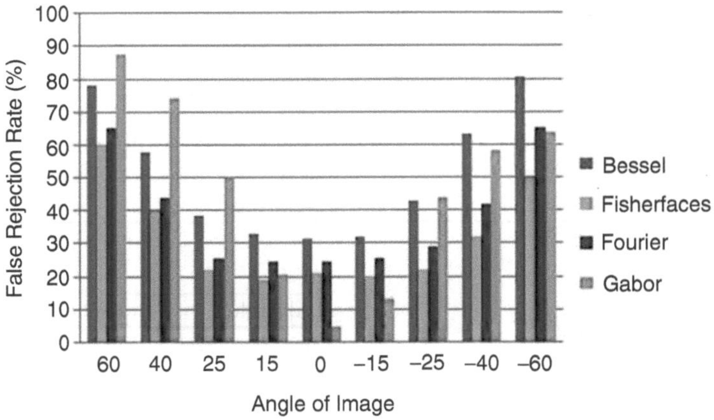

Fig. 5.4 Effect upon the FRR with varying facial orientations

Table 5.5 Facial recognition using the composite template

Algorithm	FRR (%)	FAR (%)
Fourier-Bessel	7.8	0.04
Fisherfaces	1.1	0.006
Fourier-spectra	3.8	0.02
Gabor filters	0.6	0.003

in facial orientation. Indeed, with current levels in the FRR, none of the algorithms evaluated in this experiment would be of any practical relevance.

Analysing the results from experiment 2 in more depth, it becomes apparent where the majority of errors reside. Figure 5.4 presents the FRR as the angle of the facial orientation varies. It is clear from the figure that as the angle of orientation increases an increase in the FRR is also experienced. This relationship is expected as the template is generated based upon the front-facing (zero-degree) image, which will differ in its feature vector to a larger degree the more obtuse the angle. It is interesting to note that no such relationship exists in the FAR, with the performance broadly flat across each facial orientation.

As shown in Table 5.5, using the composite template approach has significantly improved the FRR. Indeed, the FRRs are all lower than in experiment 1, demonstrating that the usability of the underlying algorithm can be improved through the use of a composite template. The important consideration is what effect this improvement has upon the level of security being provided. As shown in Table 5.5, the FARs have also improved when compared to Table 5.3. That said, care should be taken when interpreting these results. As indicated in the methodology, theory shows that the FAR in this experiment should be equal to or larger than the standard FAR (as presented in Table 5.3). In this particular experiment, this is not the case, as the type and number of verifications performed differs. Nevertheless, an important observation from these data is that the FAR has only marginally changed with an accompanying large reduction in the FRR.

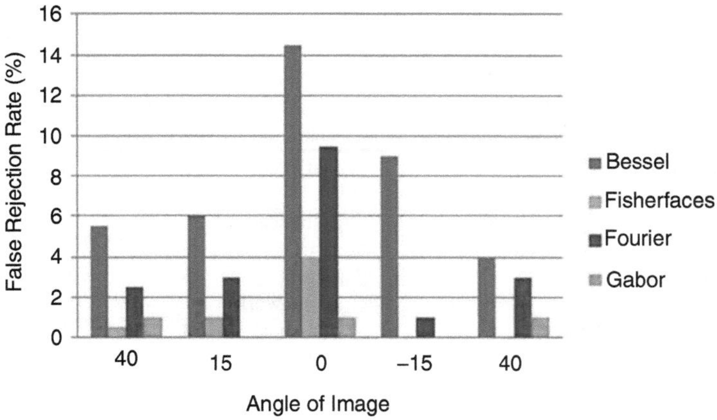

Fig. 5.5 Effect upon the FRR using a composite facial template

An analysis of the performance against the angle of facial orientation, as illustrated in Fig. 5.5, shows that the composite template is far better placed to successfully verify users with varying degrees of facial orientations. Indeed, the worst performing orientation from the results is the traditional front-facing (zero-degree) image.

The experiments presented in this study have focused upon improving the usability of facial recognition algorithms when faced with varying facial orientations – a serious issue when looking to deploy this technique transparently. Whilst other issues exist in practice, such as distance from camera and resolution of the camera, these can be addressed on an individual basis. Illumination of course will remain a problem for the foreseeable future, but this does not discount the approach from being applicable, merely that sufficient measures need to be taken when utilising the technique – for instance, within mobile device use, which represents the more challenging environment, the device could simply limit its use during daylight hours.

5.3 Keystroke Analysis

The ability to identify an individual by the way in which they type has been an interesting proposition since Spillane introduced the concept in 1975 (Spillane 1975). Since that time, a significant volume of literature has been published reporting various performance rates on differing lengths of characters. The principal concept behind keystroke analysis is the ability of the system to recognise patterns, such as characteristic rhythms, during keyboard interactions and to use these as a basis for authenticating the user. The approach utilises a keyboard as an input device – something that is present on most computing systems in some shape and form, and something users are already very accustomed to. Therefore the approach is cost-effective and good levels of acceptance exist.

Table 5.6 Summary of keystroke analysis studies

Study	Static/ Dynamic	Keystroke metrics: Inter-key	Keystroke metrics: Hold time	Classification technique	No. of participants	FAR (%)	FRR (%)
Joyce and Gupta (1990)	Static	✓		Statistical	33	0.3	16.4
Leggett and Williams (1988)	Dynamic	✓		Statistical	36	12.8	11.1
Brown and Rogers (1993)	Static	✓	✓	Neural network	25	0	12.0
Napier et al. (1995)	Dynamic	✓	✓	Statistical	24	3.8% (combined)	
Obaidat and Sadoun (1997)	Static	✓	✓	Statistical	15	0.7	1.9
				Neural network		0	0
Cho et al. (2000)	Static	✓	✓	Neural network	25	0	1
Ord and Furnell (2000)	Static	✓		Neural network	14	9.9	30

Keystroke analysis (also frequently referred to as keystroke dynamics) can be performed in static (text-dependent) or dynamic (text-independent) fashion. The former is a far simpler task than the latter. Table 5.6 presents the findings from a number of research studies that have been preformed. All, with the exception of Ord and Furnell (2000), were based upon classifying users on full keyboards, with Ord and Furnell utilising only the numerical part of the keyboard.

At first glance, it would appear that both of the dynamic-based studies have performed well against static-based approaches, given the more difficult task of classification; however, these results were obtained with users having to type up to a 100 characters before successful authentication. However, all of the studies have illustrated the potential of the technique, with Obaidat and Sadoun (1997) performing the best with a FAR and FRR of 0% using a neural network classification algorithm. In general, neural network–based algorithms can be seen to outperform the more traditional statistical approaches, and have become more popular in later studies. Notably, the original idea of keystroke analysis proposed that a person's typing rhythm is distinctive and all the original studies focused upon the keystroke latency (the time between two successive keystrokes); however, more recent studies have identified the hold time (the time between pressing and releasing a single key) as being discriminative. The most successful networks implemented a combination of both inter-key and hold-time measures, illustrating that the use of both measures has a cumulative and constructive effect upon recognition performance. Pressure has also been utilised in some studies; however, given that current hardware is not equipped with such sensors normally, this represents a deviation away from one of the key advantages of using the approach – namely its inexpensive nature as the hardware already exists.

The application of transparency to keystroke analysis is a relatively simple one from a theoretical perspective. Dynamic-based keystroke analysis is by its nature transparent, providing authentication of the user independently of any particular

Table 5.7 Performance of keystroke analysis on desktop PCs

Feature	Standard deviation	FAR (%)
Digraph	0.7	4.9
Trigraph	0.5	9.1
Keyword	0.6	15.2

string or passphrase through typing characteristics that the user exhibits whilst composing emails or writing a report. In practice, however, dynamic-based keystroke analysis can also raise a number of issues. In order to explore the application of keystroke analysis and to better understand how it functions, two studies in particular will be described that have focused upon its use in a transparent fashion. The first paper was published by Dowland and Furnell (2004) and is orientated towards desktop systems with full keyboards. Whilst other studies have also done this, this study has taken a longer-term perspective on the trial, obtaining more realistic samples. The second by Clarke and Furnell (2006) focuses upon the more restricted input characteristics of mobile phones. It represents the first study looking into the application of keystroke analysis on mobile phones and presents results for both static- and dynamic-based inputs.

The study by Dowland and Furnell (2004) collected data from 35 users over a period of 3 months. The study focused upon the capture of three forms of data:

- Digraphs – the time between two successive key presses. Also referred to in literature as the inter-keystroke latency.
- Trigraphs – the time between three successive keystrokes.
- Keywords – the time taken to type a keyword. A total of 200 of the most common English language keywords were included.

During this time over five million of the aforementioned samples were obtained. A level of pre-filtering took place, removing outliers through applying a band-pass filter (latencies between 10 and 750 mS were included). This is a standard approach in keystroke analysis studies and represents natural errors such as pressing two keys together, giving rise to a sub-10 mS response, and pauses in typing, perhaps due to distractions, causing very large latencies. The classification approach employed was a simple statistical-based technique where the mean and standard deviation of a training profile were created. Should the test samples fall within a defined range of standard deviation, the sample is accepted, otherwise rejected. It is an extremely simple form of classification, with many more sophisticated approaches available, but it does represent a worst-case result in terms of classification performance. The advantage of using such an approach is the speed at which classification decisions can be made – as only a simple comparison of two numbers is necessary, rather than a complex calculation with a neural network for instance. As shown in Table 5.7, the performance from the digraph feature proved most successful with a FAR of 4.9% and corresponding FRR of 0% (all results in the table are with respect to an FRR of 0%).

Further analysis of the data showed that on average a total of 6,390 digraphs were typed before an impostor was detected. This compares well against the 68,755 digraphs required before the authorised user was challenged. Whilst authorised

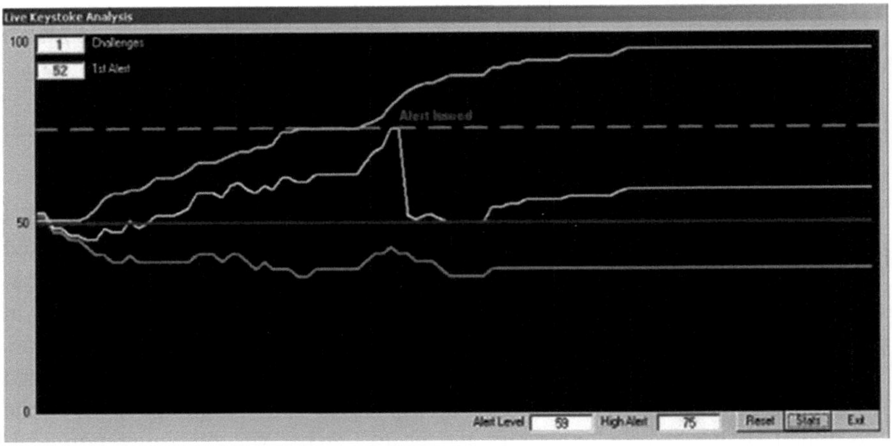

Fig. 5.6 Continuous monitor for keystroke analysis (Dowland 2004)

and impostor profiles are distinctive, it does raise a concern that an impostor would have an opportunity to type some considerable text prior to detection. It is worth highlighting that analysis of individual users shows that some have excellent levels of performance (19 out of the 35 experiencing 0% FAR and FRR) but conversely a number of users effectively being unclassifiable.

Further work by Dowland (2004) extended these initial findings into an operational prototype that continuously monitors keystroke behaviour. Figure 5.6 provides a graphical representation of the system in real time: the white line represents the continuous alert level; the green line the digraph alert level; the yellow line the trigraph alert level and the blue line the keyword alert level. Beyond a predefined threshold the system would alert – with a typical response being given to an administrator or a request made to the user to verify authenticity.

Whilst the applicability of keystroke analysis on keyboards has been established for some time, the rise of mobile computing devices, such as the mobile phone and personal digital assistant (PDA), has given rise to a variety of input devices that have a very different tactile aspect (as illustrated in Fig. 5.7). Keypads, keyboards and touch-sensitive screens are all very popular interfaces. Given the inability to type using the full range of fingers and both hands, as one can touch-type on a keyboard, questions are raised about the scope for deploying keystroke analysis on devices that incorporated a smaller set of potential characteristics. Could sufficient discriminative information be contained in a smallest of interactions? A series of initial studies by the author and colleagues sought to investigate this issue (Clarke and Furnell 2003, 2006; Karatzouni et al. 2007a).

Preliminary studies focus upon the mobile phone keypad – representing the smallest data entry device. Various aspects were tested including:

- Type of data entry – dialling a telephone number and typing a text message
- Keystroke features – inter-keystroke latency and hold time

Fig. 5.7 Varying tactile environments of mobile devices

These characteristics were selected as a direct consequence of the type of interaction being conducted. Traditionally the inter-keystroke latency has been proven to be the stronger discriminative characteristic. However, with the increase in the number of possible digraph combinations with alphabetic inputs (26 × 26), an alternative characteristic is required to reduce the degree of complexity and maximise the possible discriminative information. The use of both characteristics enables an evaluation of their individual suitability.

In order to log the data for both investigations, software was written to permit the capture of key press data – with keystroke latency and hold-time parameters being subsequently calculated upon the data. In addition, a standard PC keyboard was deemed an inappropriate means of data entry, as it differs from a mobile handset in terms of both feel and layout, with users likely to have exhibited a markedly different style when entering the data. As such, the data capture was performed using a modified mobile phone handset (Nokia 5110), interfaced to a PC through the keyboard connection.

For the telephone entry, a total of 32 participants were asked to enter a series of telephone numbers, with 30 participants taking part in the text message study (involving the entry of text messages). In each experiment, 30 samples were taken. Two thirds of these inputs were utilised in the generation of the reference profile, with the remaining used as validation samples. The pattern classification tests were performed with one user acting as the valid authorised user, whilst all the other users are acting as impostors. Any mistakes in typing made during the experiment were removed and the user was asked to re-enter the number. Again, this procedure and the subsequent evaluation are based upon a series of previous studies examining the use of keystroke analysis.

The telephone entry experiment utilised the inter-keystroke latency and was applied to static and dynamic entry. Whilst both could be used transparently, the former would take far longer to develop, as sufficient samples of each telephone number would be required so that a profile could be generated. The dynamic approach would simply require a series of telephone samples (independent of the number).

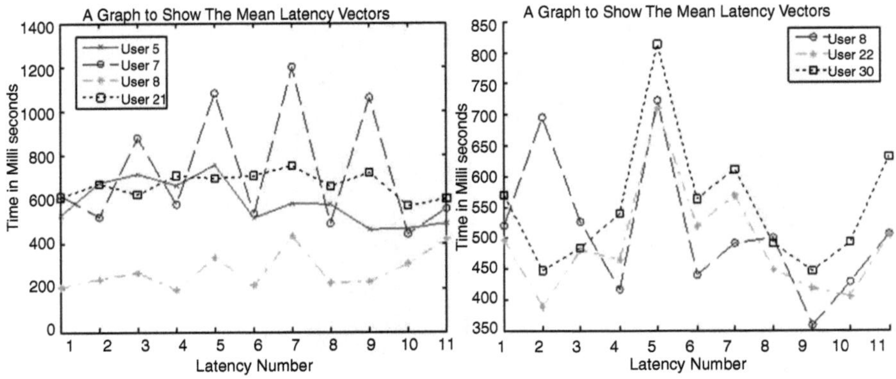

Fig. 5.8 Variance of keystroke latencies

The text-based experiment sought to evaluate the discriminative value of the hold time, primarily because with alphabetic keys the inter-keystroke latencies are increased from 100 in the numerical case to 676. By utilising the key hold time, the number of possible combinations is reduced to just 26. This helps ensure that a user would typically enter enough of the character combinations in any single text to perform authentication. In this particular study, the hold time is defined as the time taken from initial key press down to the final key press up. So on a mobile handset, generating many of the letters involves pressing a key more than once (e.g. the letter 'c' requires the number 2 button to be pressed three times). The feature vector comprises of the six most reoccurring characters (i.e. e, t, a, o, n, i).

An analysis of the input data allows an insight into the complexities of successfully authenticating a person from a single input vector of latency values. The problem is that latency vectors observed from a single user may incorporate a fairly large spread of values and as such do not exist on clearly definable classification regions. Figure 5.8 illustrates some similar and dissimilar input vectors as an indication of the complexity and subsequent difficulty the matching subsystems have in discriminating between users, the latency number on the x-axis indicating the time between keystrokes (i.e. latency number 1 corresponds to the time between the first and second keystrokes).

The results from the subsequent matching subsystems show that successfully performing keystroke analysis on mobile phones is more challenging. The results presented in Fig. 5.9 show the performance varies from an EER of 5% with the static telephone entry to an EER of 26% with the dynamic telephone entry. The three sets of figures are the result of applying three different approaches to the classification procedure:

1. Feed-forward Multi-Layered Perception (FF MLP) neural network – an identical network configuration is used for all users.
2. Gradual Training – also utilising the FF MLP but optimising the training cycles (epochs) on a per user basis.
3. Best Classifier – the results from individually selecting the most optimum network and training configuration on a per user basis.

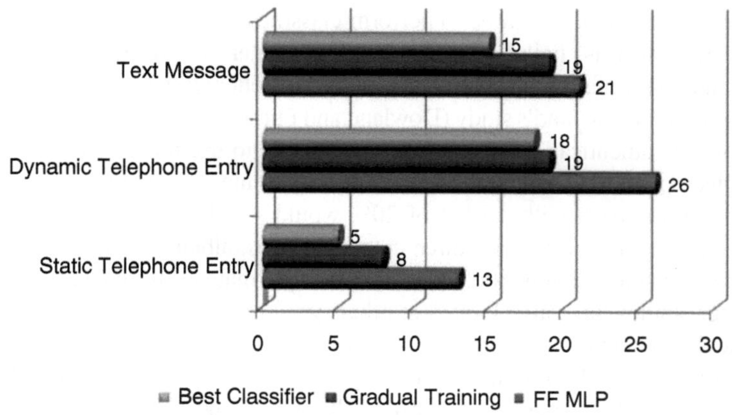

Fig. 5.9 Results of keystroke analysis on a mobile phone

Table 5.8 Keystroke analysis variance between best- and worst-case users

	Best		Worst	
Classification algorithm	User	EER (%)	User	EER (%)
Static telephone number	8	0	16	17
Dynamic telephone number	10	2	7	35
Text message	23	6	9	31

Whilst in comparison to many physiological biometrics, whose performance you would expect to be less than 5% in practice (theoretically sub-1% in many cases), keystroke analysis does appear to perform well and the ease of collection makes the approach very transparent. Moreover, these results are based upon very small entries – in comparison to the desktop studies, which obtained better results. A system that utilises more text prior to matching would perform better.

Although these average performances show promise, it is prudent to analyse the individual user performances. Table 5.8 gives the best and worst individual results experienced by users, with the static telephony entry experiencing the lowest performance of 0% with user 8. However, it is also noticeable that the performance goes as high as 35% for user 7 in the dynamic telephone experiment.

The investigation has shown the ability for classification algorithms to correctly discriminate between the majority of users with a relatively good degree of accuracy based on both the inter-keystroke latency and the hold time. The hold time is an unusual keystroke characteristic to use on its own, but has proved useful in this investigation as it avoided the problem of sampling and profiling the large number of digraph pair combinations that would have usually been required. Authentication performance could, however, be increased if the classification algorithm utilised a number of techniques to classify the subscriber, capitalising on the specific content of the message. For instance, in a worst-case scenario, the hold-time classification algorithms presented could be used on messages with dynamic content utilising between 2 and 6 of the most commonly recurrent characters. However, the next stage, depending on content, would be to perform classification on commonly reoccurring static words, such as 'hello', 'meeting' and 'c u later' where both inter-keystroke latency and hold-time characteristics can be used to better classify the subscriber, such as was utilised in Dowland's study (Dowland and Furnell 2004). Given the transparent nature of the authentication, it would also be viable to use more than one approach over a given period of time, with the system responding on the aggregate result. For instance, an approach with a FAR of 20% would see this reduce to 4% with two authentications and 0.16% with three authentications, albeit with an increase in the FRR. The decision need only be taken after multiple matches have taken place.

Further studies have also looked at other types of interface, such as the thumb-based keyboards and touch-sensitive keyboards (Karatzouni et al. 2007a; Saevanee and Bhattarakosol 2008). In all, the levels of performance reported are in line with the previous results presented, suggesting that whilst the differing tactile environment certainly plays a role in producing a different set of features, the different nature of the form factor also provides a source of discriminative information in its own right, leading to positive results. Unlike other biometrics, however, each feature profile is unique to the tactile environment (i.e. keypad, thumb-based keyboard or full-size keyboard) requiring independent enrolment on each type of device.

A point identified in a number of studies is the failure of keystroke analysis to perform successfully for a minority of users. These users tend to have high inter-sample variances and exhibit few distinctive typing rhythms. As such, any authentication system that implements a keystroke analysis technique would also have to consider the small number of users that will experience too high an error rate in order to ensure both the security and user convenience factors required by the overall system are met. It is not a solution for all – but few biometric approaches even used in the traditional point-of-entry format are.

5.4 Handwriting Recognition

Handwriting recognition refers to the ability to identify a person by their handwriting. In the general sense, this has referred to the examination of hand-written materials for use within forensic applications. It is a semi-automated process that typically examines a piece of written material and provides a list of potential candidates. It has not been applied to authentication. Similar to signature recognition, static and dynamic approaches exist. The former looks at the shape of letters, slopes and distance between words. The latter can be achieved through determining speed, flow, pauses and possibly even pressure.

With the increased use of touch-sensitive screens – mobile phones, PDAs, tablet PCs and increasingly desktop systems – interaction through handwriting rather than keyboards is becoming more popular. Whilst this approach has little use within point-of-entry authentication – where signature recognition would be the natural choice – for transparent authentication, handwriting recognition could play a vital role in authenticating users whilst they are taking notes.

As no fully automated systems for use within an authentication system exist, it clearly becomes difficult to establish to what degree this approach would function within this context. In order to obtain an idea (worst case) of the performance and operation of handwriting recognition within this context, previous research has focused upon the application of signature recognition to the problem (Clarke and Mekala 2007), where the signature is replaced by a word. The signature recognition approach contains all the necessary capture, processing, extraction and matching subsystems required for authentication. The question is whether sufficient discriminatory information is contained within the same written word that everyone can write.

An experiment was devised to examine the effectiveness of signature recognition applied to handwriting verification. The experiment consisted of two parts:

- A control experiment to test the technique under normal conditions (i.e. determine verification performance based upon signature entry)
- A feasibility experiment that replaced the signature input with a series of commonly used words

Commercially available signature recognition software for PDAs was used and 20 participants took part in the experiment (Turner 2001).[1] PDALok utilised the dynamic rather than static approach to classification. The performance of the control experiment was excellent, with a FAR of 0% and a good FRR of 3.5%.

Surprisingly, however, the performance of the feasibility experiment surpassed that of the control experiment, performing on average with a FAR and FRR of 0% and 1.2% respectively. With a more detailed examination of the results, Fig. 5.10 gives the average FRR across all eight words for all the participants in the feasibility experiment. Thirteen participants achieved 0% FAR and FRR, with user 8 performing worst with an FRR of 8.8%.

Analysing the performance of the signature recognition technology against each of the words, as shown in Table 5.9, shows that the length of the word was not a factor in security, as the technique can already provide an adequate level, but rather has an effect on the FRR and subsequent usability of the approach, with longer words having a greater FRR. From this result it is suggested that sufficient discriminatory information is contained within small words without requiring a user to sign longer words.

Weaknesses of the study obviously relate to the population size of the participants. Although a larger population of participants would have been ideal, the overall performance results were based on 8,298 comparisons. Furthermore, the experiment only included zero-effort attacks rather than adversary attacks, the latter approach referring to a technique where attackers purposefully look to copy the sample (e.g. attempting to forge a sample after having access to the original). Whilst present in all biometric techniques, its use within signature recognition has been a long-established attack vector. That said, the initial findings would suggest handwriting recognition for authentication has a valuable contribution to make in future transparent authentication.

[1] Unfortunately it appears this software is no longer available.

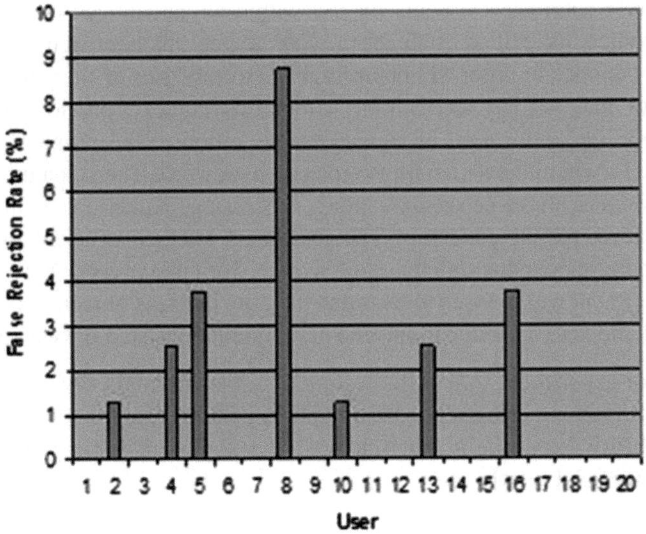

Fig. 5.10 Handwriting recognition: user performance

Table 5.9 Handwriting recognition: individual word performance

Word #	Word	FRR (%)	FAR (%)
1	Bye	0	0
2	Love	0.5	0
3	Hello	1	0
4	Sorry	0	0
5	Meeting	1.5	0
6	Thank you	2.5	0
7	Beautiful	1.5	0
8	Congratulations	2.5	0
	Average	**1.19**	**0**

5.5 Speaker Recognition

For telephony-based applications, speaker recognition is the obvious approach to utilise. With the widespread use of mobile phones and increasing use of Voice over IP (VoIP) applications such as Skype, voice-based communications are an immense source of information. Indeed, given the potential for ease of use and the lower cost (due to hardware already being deployed), speaker recognition has been the focus of a great amount of research over the years. It can enable call centres to authenticate customers, and provide law enforcement and the intelligence services the opportunity to identify individuals from telephone calls. Unfortunately, however, the performance of speaker recognition has tended to fall short of expectations.

The environment, in particular the surrounding noise, and the microphone are two significant problems that affect performance.

Speaker recognition can be achieved in effectively one of two modes: text-dependent and text-independent. From a transparent authentication perspective, it is the latter that would provide the opportunity to remove the intrusive nature of authentication. It is, however, also the more difficult problem to solve. Text-dependent systems outperform their text-independent counterparts with lower error rates. Whilst this is likely to remain true for the foreseeable future, significant more research is being undertaken within the text-independent domain – for obvious national security reasons. Research into text-independent speaker recognition has been ongoing for over 40 years. Since 1996, the National Institute of Standards and Technology (NIST) has been coordinating Speaker Recognition Evaluations (SRE), which have provided the community with a uniform set of data from which to evaluate and compare their algorithms.

Text-independent speaker recognition systems are broadly based upon two approaches: spectral-based systems and higher-level systems. The former are the more established approaches but the latter are relatively new and are providing more scope for further research. Short-term spectral analysis is used to model the different 'sounds' a person can produce, the reasons for the different 'sounds' being based upon the physical vocal tract and associative sound production equipment of the individual (Jain et al. 2008). It is based upon an acoustic examination of the signal. Within this area, amongst the most successful techniques is the Gaussian Mixture Model – where a mixture of multidimensional Gaussians tries to model the underlying distribution (Reynolds et al. 2000).

Higher-level approaches tend to look for speaker idiosyncrasies, also referred to as 'familiar-speaker' differences (Doddington 2001). This approach looks at the longer-term usage of words or phrases and to the features that are associated to them – such as inflections, stress and emphasis. For instance, bigrams such as 'it were', 'you bet', 'uh-huh' and 'you know' are the result of the individual rather than necessarily the context. These unique aspects of the way in which individuals use language can be used to assist identifying individuals. Rather than being a function of our physical vocal tract, these higher-level approaches focus more on our learnt behaviour and the environment in which the user lives. The study by Donnington (2001) involved utilising a dataset known as Switchboard. The dataset comprises 520 participants. The experiment was conducted under the same methodology as described by NIST's extended speaker detection test (NIST 2011) and obtained an (approximate) EER of 9% (after pruning bigrams that appear less than 200 times from the dataset).

Suffice to suggest that speaker verification approaches are invaluable for use within transparent authentication systems (TASs). Whilst their performance has fallen short for many applications, compared to many biometric techniques that are applicable in this context, speaker verification is amongst the better performing approaches. When applied to a multi-modal or composite authentication system, the approach provides a useful identity check when few other biometric techniques would be appropriate.

5.6 Behavioural Profiling

Behavioural profiling is an interesting proposition from a transparent authentication perspective. Through modelling of a user's interaction with a computer, behavioural profiling is a natural approach to the problem. Much of the initial research in this domain was not specifically linked to authentication but to fraud detection. A number of studies throughout the 1990s focused upon developing network-based monitoring systems for large telecommunications companies and financial institutions (trying to resolve mobile telephone and credit card fraud respectively). More recently, increased focus has been attributed to the use of such approaches within authentication and intrusion detection systems (IDSs) of information technology (IT) systems, the use of the approach within IDSs being applicable because of the transparent capturing of samples – a deviation away from normal behaviour providing an alert to the anomaly-based IDS. Indeed, there are close analogies (but different aims) between a transparent authentication system and an anomaly-based IDS.

Whilst usability and acceptability criteria are good, the approach suffers from issues related to the uniqueness and permanence – the latter being a serious problem. Whilst research (presented shortly) has indicated that people do present a level of uniqueness with respect to their interactions, the nature of those interactions varies with time, resulting in particular interactions only being present for a relatively short period of time. This presents a problem to the biometric designer from a template generation and aging perspective. It is imperative to maintain a template that is relevant to the user but without introducing impostor data. Pattern classification algorithms utilised within the matching subsystem tend to be split into two approaches: supervised and unsupervised. Both approaches refer to the nature of training that takes place within the classification algorithm. Supervised approaches utilise historical data with known authorised and impostor data flagged. In this manner, the algorithm is able to specifically learn to differentiate between the two datasets. Unsupervised learning utilises the same historical data, but no indication of impostor versus authorised data is given. That is for the unsupervised algorithm to determine. The unsupervised approach is obviously the more challenging classification problem to solve; however, it circumvents the single biggest problem with behavioural profiling, the misclassification of the legitimate user samples as valid rather than as impostors.

Given the limited commercial exploitation of behavioural profiling, it is useful to explore the research into behavioural profiling in both the fraud and authentication domains. Within the area of fraud detection, one of the more significant initial studies was undertaken by the European Advanced Security for Personal Communications (ASPeCT) project (Gosset 1998), which sought to develop a fraud detection system for mobile communications. The types of features used to create the user profiles were gathered from a toll ticket (TT) – a bill issued by the network after each call, and consisted of the following features (Moreau et al. 1997):

- International Mobile Subscriber Identity (IMSI)
- Start date of call
- Start time of call

Table 5.10 ASPeCT performance comparison of classification approaches

Classification approach	Detection rate (%)	False alarm rate (%)
Supervised	90 (60)	3 (0.3)
Rule-based	99 (84)	24 (0.1)
Unsupervised[a]	64	5

Brackets indicate more practical performance
[a]Estimated from results graph

- Duration of call
- Dialled telephone number
- National/International call

These features were constantly updated after each call and used to maintain a *Current Behaviour Profile* (CBP), which comprises individual user profiles of the most recent activity. Differential analysis enables the individual profile to be taken into account and assists in compensating against the general population analysis. It compares the CBP or short-term activity with the *Behaviour Profile History* (BPH) or long-term activity. An initial evaluation of three classification approaches was undertaken, utilising an identical methodology using real data (Stormann 1997). As shown in Table 5.10, the unsupervised approach performed the worst; however, given the lack of classified training data, this would be expected. The results from both the rule-based and supervised approaches were significantly better.

The study authors conclude that the most promising development is to combine the three classifiers into a hybrid tool, referred to as BRUTUS (B-number and RUle-based analysis of Toll tickets utilising Unsupervised and Supervised neural network technologies) (Gosset 1998). Through the use of multiple classifiers it is suggested that strengths of each approach can be capitalised upon:

- Unsupervised neural network for novelty detection
- Supervised neural network for detecting when user behaviour is similar to previously observed and fraudulent data
- Rule-based analysis for explaining why alarms have been raised and subsequently to develop extra rules

Unfortunately, results on how effective this combined classifier is have not been published. Nevertheless, the results from use of the classifiers on an individual basis are promising. However, the use of multiple classifiers within a single overall system is an increasingly popular approach. Studies by Provost and Aronis (1996), Chan et al. (1999), Stolfo et al. (2000) and Shin et al. (2000) have all proposed multi-classifier approaches.

Much like the concept behind transparent authentication – to enable commensurate security linked directly with what the user's activity is rather than overburdening the user when undertaking less sensitive operations – researchers also realised that not all fraudulent activity was equal. More effective detection systems could be created if an understanding of the impact could be devised. Stolfo et al. (2000) in particular published a paper specifically focusing the detection engine towards minimising cost.

Table 5.11 Cost-based performance

Type of classification	Number of classifiers	Accuracy (%)	Savings ($)
Commercial off-the-shelf product	–	85.7	682k
Best base classifier	1	88.8	840k
Meta-classifier	50	89.6	818k

Indeed, Stolfo et al. state that 'traditional methods used to train and evaluate the performance of learning systems when applied to fraud and intrusion detection are misleading or inappropriate as they provide an indication of the absolute number of fraudulent or intrusive actions but take no consideration as to the individual value of the activity'. For instance, the performance of a system detecting 10,000 fraudulent activities might appear to outperform a system that merely detects 100. However, should the value of the detection in the latter case be $100 versus $0.5 for the former, it is obvious that the latter system performs better with respect to the cost and the former with respect to the volume of detections. This raises an important aspect in the design of the classifier: what is the classifier specifically designed to do. It is essential to design the classifier to meet the exact requirements of the system. For instance, the following three definitions would all result in a differing classifier:

- To detect fraudulent activity
- To detect fraudulent activity according to cost
- To detect fraudulent activity that is cost effective to follow up

The first classifier would simply be designed to alert any and all fraudulent activity; the second would utilise cost-based modelling to base alarm decisions upon and the third would only raise alarms on activity that would exceed a user-defined threshold value.

Stolfo et al. (2000) developed a cost model for credit card fraud based upon the sum and average loss caused by fraud. Interestingly, at a higher level, their approach was less concerned with the actual classifier, but rather with the evaluation of 'base classifiers' (a variety of classifiers) that sought to maximise total cost savings. Utilising real data of fraudulent activity with an 80% non-fraudulent and 20% fraudulent distribution, the authors evaluated their approach against a commercial off-the-shelf product the bank was already utilising. As shown in Table 5.11, the cost-based model using the *best base* classifier did improve accuracy and overall savings. Interesting to note, a multi-classifier approach utilising all 50 of the base classifiers improved the accuracy results still further but not the savings. The authors put this down to the base classifiers still being biased towards minimising statistical misclassifications rather than the cost model.

From the supervised learning section a number of relevant concepts are presented which are particularly relevant to the problem:

1. The use of multiple classifiers
2. The use of absolute and differential analysis of data
3. Dynamic creation and pruning of rule sets
4. Cost-based modelling
5. Dynamic selection of classifiers based upon the individual problem

From an analysis of the performance of supervised approaches, all techniques outperform unsupervised approaches and the majority of them perform to a level that would be operationally useful and cost effective. With a well-defined problem and training data that is truly representative of the problem, supervised classifiers are very effective.

However, it is important that care be taken in the design of supervised classifiers; for instance, unbalanced class sizes for training (where the legitimate transactions far exceed fraudulent cases) can cause an imbalance and misspecification of the classification model (Bolton and Hand 2002). It is also imperative to realise that supervised approaches will never directly adapt to new types of fraudulent events. The process requires human understanding of the event prior to its inclusion within the classifier. Given the dynamic and adaptable environment of misuse detection, this will always leave the network open to new forms of attack for a period of time.

Unsupervised approaches are not appropriate to all classification problems, particularly those with well-defined classifiable objects. It is the property of detecting previously unknown forms of fraud that make unsupervised learning approaches of particular interest in this problem. However, due to the performance constraints imposed by unsupervised techniques, many approaches to date have been confined to acting as pre-processing mechanisms, such as filtering out normal data to reduce the load for a subsequent supervised classifier or rule-based system (Kou et al. 2004).

Within the realm of authentication, researchers have begun to look at the role of behavioural profiling as a mechanism for providing identity verification, but the volume of research is significantly reduced when compared to fraud detection. A preliminary study undertaken by the author sought to understand to what degree people exhibited unique behaviour whilst using their desktop computer (Aupy and Clarke 2005). The study involved a relatively small cohort of 21 participants over a capture period of 60 days. There are also some severe limitations to the study; however, it does help demonstrate several interesting aspects with respect to template aging and the possible viability of the approach in a desktop environment. To enable data collection an application was developed that was capable of capturing system events so a profile of user activity could be built. A front-end application was developed to extract the data from the database, as illustrated in Fig. 5.11.

An issue that became evident quickly within the research was the volume of data being captured and the resulting size of the database and storage required. As such, novel mechanisms were developed that simply stored the metadata with respect to interactions in a compressed fashion. The key features captured are presented in Table 5.12.

From a pattern classification perspective, the study utilised a neural network configuration known as the Feed Forward Multi-Layered Perceptron (FF-MLP). FF MLP networks have particularly good pattern-associative properties and provide the ability to solve complex non-linear problems (Bishop 2007). In addition to the features identified above, a timestamp was also included. The timestamps were divided into quarters of an hour, giving rise to a range of input values of 0–95 during a day. Preliminary analysis found that the four features alone were not sufficient for

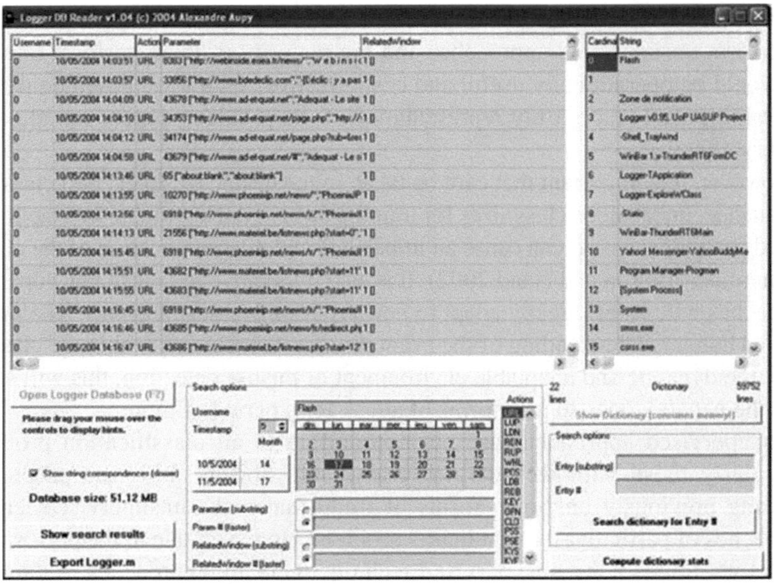

Fig. 5.11 Data extraction software

Table 5.12 Behavioural profiling features

Code	This action is raised each time...
KEY	A full word has been typed in. The word is recorded, along with the title of the window where it has been entered.
OPN	A window is opened. The name and class of the window are recorded.
CLO	A window is closed. The name and class of the window are recorded.

successful classification. As such, it was decided to utilise 300 repetitions of the aforementioned features, giving rise to an input feature vector of 1,200. On average these 300 actions corresponded to approximately 10 min of user interactivity.

The collection period of 60 days was split into four 2-week chunks, with each user taking the turn of the authorised user with the remaining acting as impostors. Due to the quantity of input data, the networks could only be designed with two users – an authorised user and an impostor – creating a total of 441 networks. Although this is not ideal, for the purposes of a feasibility study it will permit an insight into whether Service Utilisation Profiling is plausible.

The calculation of the error rates and subsequent performance is also based upon a majority voting technique, where one classification is based upon three network outputs. Two passes and a fail equal a pass and two fails and a pass corresponds to a fail. This effectively results in the system being able to provide an authentication decision for every 30 min of user activity. The results from the study are encouraging, with typical EERs of below 10% and an overall average EER of 7.1%, as shown in Table 5.13.

Table 5.13 Behavioural profiling performance on a desktop PC

User	EER (%)	User	EER (%)
1	3.8	12	6.9
2	9.4	13	5.0
3	7.6	14	5.1
4	3.8	15	10.7
5	8.5	16	3.0
6	7.6	17	6.2
7	6.0	18	6.0
8	6.1	19	7.4
9	5.4	20	6.8
10	10.5	21	6.2
11	6.7	**Average**	**7.1**

Some care must be taken with these figures, however, as they arguably overestimate the probable performance one would expect in practice for two reasons:

1. The networks are generated and tested against one impostor user each time and not against all impostors (as would be typical). In practice, a single network would have to stop all impostors.
2. The EERs are calculated based upon the complete 60 days of input data. The first fortnight of data, however, is also utilised in the training of the network. You would therefore expect the network to perform well against these data.

Nonetheless, the results are still promising given the relatively small amount of information/interactions that the network is utilising. Further analysis of these problems with the performance shows that they are not substantial problems. Although training a network with a single impostor is not normal procedure, unfortunately processing constraints restricted using more impostor data. However, tests performed where a network was trained using one authorised user and an impostor, but validated using another impostors' data, have shown encouraging results. User 6 was trained using User 7 data as impostor, but validated using Users 10 and 15 and achieved an EER of 0%.

Figure 5.12 illustrates the output from the neural network given legitimate and impostor data, split into fortnightly sections. Although the second performance problem certainly has the effect of improving the FAR and FRR, it can be seen from Fig. 5.12 that network performance is still strong given the second fortnight of data, which has not been previously shown to the network. This is one of the primary indicators to the success of behavioural profiling. What can also be seen from this figure is how the network output becomes noisier as time progresses – illustrating the issue of template aging that was highlighted at the beginning of this section. However, even in the fourth quarter it can be seen that the majority of impostor data still resides around zero and with a carefully selected threshold level authentication can still be successful.

Further research has also been undertaken on the development of behavioural profiling for mobile devices. Whilst the aforementioned research into fraud detection

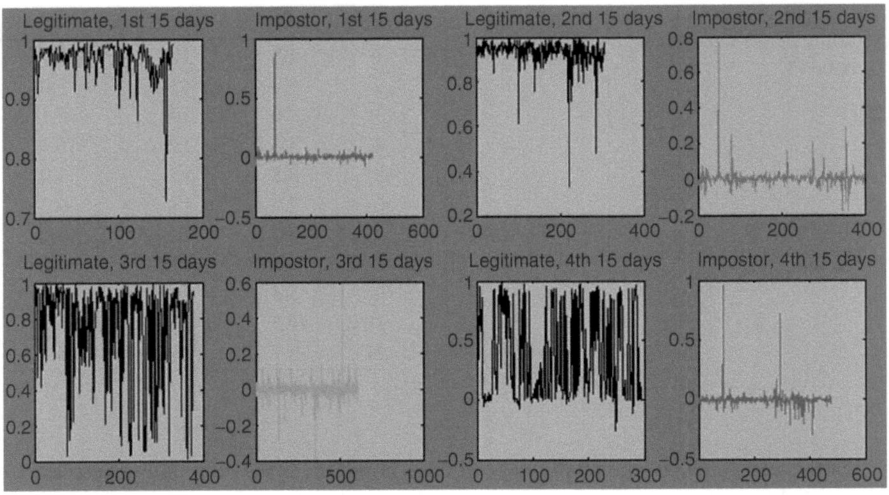

Fig. 5.12 Variation in behavioural profiling performance over time

provided a network-based classification of misuse, more recent research has focused upon host-based detection. Given the advancement in technology and functionality, mobile devices are no longer serviced by just one network but are able to connect to a variety of networks (including Wi-Fi networks operated by different providers and Bluetooth networks). As such, from an authentication perspective, network-based solutions that are connected to a particular service provider would only serve to capture a subset of possible misuse. Research carried out by Li et al. (2011) has demonstrated some very promising results (without the experimental limitations that the previous desktop PC presented).

The experiment employed a publicly available dataset provided by the MIT Reality Mining project (Eagle et al. 2009). The dataset contains 106 participants' mobile phone activities from September 2004 to June 2005. By using preinstalled logging software, various mobile data were collected from participants' Nokia 6600 mobile phones and an overview of the data is presented in Table 5.14. The MIT Reality dataset contains a large and varied selection of information, which covers two levels of application usage: application-level applications (general applications) and application-specific applications (voice call and text message).

By default, a number of common applications are preinstalled on the mobile device by the manufacturer, such as phonebook, clock and voice calling. With increased computing processing power and storage space and almost 15,000 new mobile applications becoming available on the market every month, mobile users have the freedom of installing any additional applications on the device (Distimo 2010). From a high-level perspective, the general use of applications can provide a basic level of information on how the mobile user utilises the device. Such information could be the name of the application, time and location of usage. Applications themselves also provide a further richer source of information. Within many applications the user connects to data that could provide additional discriminatory

Table 5.14 MIT dataset

Activity	Number of logs	Information contains
General applications	662,393	Application name, date, time of usage and cell ID
Voice call	54,440	Date, time, number of calling, duration and cell ID
Text message	5,607	Date, time, number of texting and cell ID
Bluetooth scanning	1,994,186	Date, time of each scan and individual device's MAC address

Table 5.15 Application-level performance

		Number of log entries (%)					
		1	2	3	4	5	6
Profile technique	Static 14 days	21.1	17.4	16.3	14.9	14.2	13.6
	Dynamic 14 days	21.1	17.3	16.0	14.5	13.9	13.5
	Dynamic 10 days	22.1	17.8	16.2	14.6	14.4	13.7
	Dynamic 7 days	24.0	19.4	17.6	15.9	15.3	14.4

information. For instance, when surfing the Internet, the Internet browser will capture all the universal resource locators (URLs) an individual accesses. The research sought to evaluate both application-level and application-specific behavioural profiling.

Two types of profile techniques were employed: static and dynamic. For the static profiling, each individual dataset was divided into two halves: the first half was used for building the profile, and the second half was used for testing. For the dynamic profiling, the profile contains 7/10/14 days of user's most recent activities. At no point is any of the training dataset used to validate the performance. A smoothing function that considers a number of entries is also employed. The basis for this approach was derived from the descriptive statistics produced when analysing the data and the large variances observed. A dynamic approach attempts to mitigate against the template-aging problem.

For general applications, the following features were extracted from the dataset: application name, date of initiation and location of usage. The final dataset utilised contained 101 individual applications with 30,428 entry logs for 76 participants. Among these 101 applications, the phonebook, call logs and camera are used by all participants. By using a simple formula, a final set of equal error rate (EER) for users' application-level application usage is presented in Table 5.15. The best EER achieved is 13.5%. It was obtained by using the dynamic profile technique with 14 days of user activity with 6 log entries. In comparison, the worst performance was achieved by using the dynamic profile technique with 7 days of user activity with 1 log entry.

For the telephone call application, 71 participants used it during the chosen period, with 2,317 unique telephone numbers used and 13,719 call logs. The following features were chosen for each log: the telephone number, date and location of calling. The best experimental result is an EER of 5.4% and it was achieved by using the dynamic profile technique with user's most recent 14 days' activity and 6 log entries. In contrast, the worst result is almost double the best performance, and it was obtained by the configuration that used the dynamic 7 days template with 1 log entry (Table 5.16).

Table 5.16 Application-specific performance: telephone app

		Number of log entries (%)					
		1	2	3	4	5	6
Profile technique	Static 14 days	9.6	9.1	7.9	7.2	4.3	6.4
	Dynamic 14 days	8.8	8.1	6.4	6.4	6.3	5.4
	Dynamic 10 days	9.6	8.6	8.1	7.2	6.9	6.0
	Dynamic 7 days	10.4	8.8	8.5	7.3	7.0	6.2

Table 5.17 Application-specific performance: text app

		Number of log entries (%)		
		1	2	3
Profile technique	Static 14 days	7.0	4.3	3.6
	Dynamic 14 days	5.7	2.6	2.2
	Dynamic 10 days	8.3	4.1	3.7
	Dynamic 7 days	10.7	5.7	3.8

For the text messaging experiment, 22 users' text messaging activities were available during the chosen period. The text messaging dataset contains 1,382 logs and 258 unique telephone numbers. For each text log, the following features were extracted: receiver's telephone number, date and location of texting. Due to certain participants having limited numbers of text messaging logs, a maximum of 3 log entries were treated as one incident. The final result for users' text messaging application is shown in Table 5.17, with the best EER of 2.2%. It was acquired by utilising the dynamic profile method with 14 days of user activity and 3 log entries. The smoothing approach plays a significant role in improving the performance, with an almost halving of the performance between 1 and 2 log entries.

The application name and location have proved to be valuable features that provide sufficient discriminatory information to be useful in authentication, with location being the more influential characteristic. However, whilst this might identify many misuse scenarios, it would not necessarily identify all cases of misuse – particularly those where a colleague might temporarily misuse your device as the location information is likely to fall within the same profile as the authorised user. So care is required in interpreting these results. The intra-application approach should also help to specifically identify this type of misuse, as the telephone numbers dialled, text messages sent or Internet sites visited are likely to differ from the authorised user.

In general, dynamic profiling achieved a slightly better performance than the static profiling did. This is reasonable as a dynamic profile contains a user's most recent activities; hence it obtained a more accurate detection. Furthermore, with a longer training set period, the performance also improved. Moreover, storage issues and processing issues should also be taken into consideration with larger training. While a smoothing function treated more log entries as one incident, the performance also improved accordingly. The smoothing function reduces the impact any single event might have and seeks to take a more holistic approach towards monitoring for misuse. The disadvantage of this approach is that it takes a longer time for the system to make a decision; hence, an intruder could have more opportunities to abuse a system and a certain amount of abuse could be missed by the security control. Limitations in the dataset are also likely to have created certain difficulties. As the dataset was collected in 2004, the number of mobile applications available for users to choose was limited; this resulted in a large similarity of application-level application usage between mobile users and an increased difficulty for the matching subsystem. In contrast, in the early part of 2010, there were around 200,000 mobile applications available (Distimo 2010). As mobile users have more options, their application-level usage would arguably differ more. Therefore, it would be easier to discriminate mobile users through their application-level usage.

The ability to utilise behavioural profiling for network-based detection or for local PC and mobile devices has been presented. Whilst no commercial solutions exist outside of the domain of fraud, research studies have provided a good basis for suggesting such approaches do have the ability to correctly classify users. Whilst template aging is a significant issue when looking to deploy the approach on its own, the issues can be mitigated within a multi-modal or composite authentication system.

5.7 Acoustic Ear Recognition

This particular approach is amongst the less well-established techniques. Indeed, little literature exists describing the approach and little to no empirical research has actually been undertaken to prove its viability. From a transparent authentication perspective, however, it offers some very real potential benefits and as such will be explored further.

Whilst ear geometry is concerned with the physical shape and size of particular features of the ear, acoustic ear recognition, however, is concerned about the characteristics of the inner ear, the ear canal itself. It is suggested that anatomically people have different shapes and sizes of ear channel that will subsequently cause differing levels of reflection and absorption. By playing a sound into the ear and recording the signal that is returned, the returned signal will have been modified by these unique properties. An illustration of the process is presented in Fig. 5.13. The concept of acoustic ear recognition is covered initially by a US patent filled in 1996 and more recently by a UK patent filled in 2001 (US 5,787,187 1998; GB 2,375,205 2002).

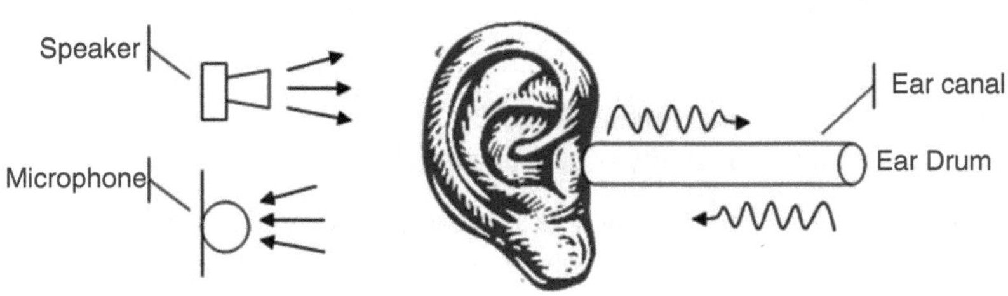

Fig. 5.13 Acoustic ear recognition

Table 5.18 Performance of acoustic ear recognition with varying frequency

Frequency (Hz)	Headphone (%)	Earphone (%)	Mobile phone (%)
1.5–22k	0.8	1	5.5
1.5–10k	0.8	1.4	6.5
10–22k	2.5	2.5	10
16–22k	8	6	18

For telephony-based applications the approach offers transparent and continuous identity verification, with an inaudible signal being transmitted in the ear during a telephone conversation. Given the challenges in speaker recognition, acoustic ear recognition offers a far more user-friendly approach to authentication. Unlike speaker recognition, it would, however, require an additional microphone sensor to be included on, near or in the ear to capture the resulting signal. Whilst this obviously has a significant impact upon its practical use immediately (i.e. the technology deployed within the market is not equipped which such a sensor), the low cost of microphone sensors would not prohibit its inclusion in the future.

A couple of preliminary studies have been published on ear acoustic recognition. Akkermans et al. (2005) published a study looking into the performance of acoustic ear recognition and compared a number of different devices/headphones to determine the effect, if any, they had (a earphone, headphone and mobile phone). The study involved 31 participants for the earphone and headphones with 8 samples of each taken. The mobile phone study involved 17 participants, again with 8 samples. Using a 5:3 split between the 8 samples and applying Fisher Linear Discriminate Analysis, the EERs for the headphone, earphone and mobile phone were 1.4%, 1.9% and 7.2% respectively. With the mobile phone performing the worst of the three hardware scenarios, it illustrates the effects and additional noise that is introduced from having the microphone further from the ear. Further analysis of the results over varying frequency ranges illustrates that the best classification performance is achieved with a wider frequency range (as shown in Table 5.18). It also demonstrates reasonable performance at higher frequency ranges, allowing for the possibility of using inaudible signals such as ultrasonic, resulting in an improvement in acceptability.

The authors of the UK patent have also published some experimental studies on ear acoustic recognition; however, this initial implementation is different from the initial concept presented in the patents (Rodwell et al. 2007). The authors suggest acoustic ear recognition can be performed in one of three ways:

1. Burst mode – in-band stimuli
2. Continuous mode – out-of-band stimuli
3. Continuous mode – naturally occurring stimuli

The generally accepted definition of acoustic ear recognition (and that defined by the patents) covers the first and second approach. The first approach refers to a computer-generated signal that is within the audible hearing range – such as noise made by a dialling tone. The second approach is to utilise an inaudible signal that is transmitted, so that the user is not inconvenienced. In this way, the signal can be transmitted continuously in order to verify the user's identity, without the knowledge or interaction of the user. The third approach refers to using naturally occurring sounds, such as those made by the individual speaking during the telephone conversation. In that way, the user is not inconvenienced with computer-generated signals. The experimental findings published have implemented the third approach. In this approach, sound is captured at both the mouth and the ear. It is suggested that the sound waves are modified by throat, nose and ear canals. A comparison of both samples provides a basis for understanding the unique characteristics of the head – not simply the ear canal (as the previous study focused upon). As such, the authors refer to the approach as the Head Authentication Technique (HAT). The feature extraction process is based upon identifying and extracting absorption levels that exist between the two signals. The study was formed from 20 participants providing 20 samples, split over two sessions. The result performance obtained using a neural network–based classifier was a false non-match rate (FNMR) of 6% and a false match rate (FMR) of 0.025%.

Both of the preliminary studies have utilised small population samples that minimise the statistical relevance of their findings. Nevertheless, both studies have demonstrated levels of performance well within the boundaries of other behavioural biometrics. Whilst the limitations of the research to date do not provide conclusive evidence as to the level of permanence, acceptability and circumventability, it is likely that the technique will perform well against each criterion. Medical studies suggest that whilst actual hearing is impacted by age, the shape and structure of the ear canal change little with time, providing a good level of template permanence. If the signal is transmitted within the inaudible frequency spectrum, acceptability issues will be effectively non-existent (unless such signals raise medical concerns amongst users). Finally, the ability to circumvent the approach is challenging giving the inability to acquire a physical representation of the trait. Much like Vascular Pattern Recognition (VPR), inaccessibility of the underlying trait provides robust protection against forgery. The technique also has the opportunity of providing truly non-intrusive authentication.

5.8 RFID: Contactless Tokens

Token-based approaches do not typically exhibit the characteristics required for transparent authentication. They rely upon the user to bring both the token and the device together when authentication is required. Within the context of the mobile phone, this subsequently resulted in the Subscriber Identity Module (SIM) card being left permanently in situ, thus removing any authentication security it could provide. Whilst contact-based tokens have an obviously intrusive aspect to their implementation, contactless tokens offer the opportunity to provide authentication transparently if implemented correctly. Carrying a contactless token in a pocket, for instance, could provide a very localised area from which to operate the mobile phone. It does not solve the remaining two issues with regard to token-based authentication: requiring the user to remember to take the token and the assumption that the authorised user has the token. However, further implementation considerations can help to mitigate these issues. For example, the contactless token could be integrated into a watch or piece of jewellery – something a user would naturally wear on a regular basis.

Contactless token technologies, such as RFID, are becoming more widely utilised in a variety of applications, from supply-chain management to travel cards. As such, they come in a variety of form factors and operational modes that suit particular applications. However, all RFID systems share three key components:

- RFID tag (or transponder) – the token that represents the identity of the object to which it is tied
- RFID reader (or transceiver) – the device that is able to communicate with the RFID tag
- Backend data processing system – to perform the necessary business logic and provide database storage

RFID-based systems operate in one of three modes, depending upon the presence and location of a power source:

- Passive – no power supply present. The required energy to power the tag and communicate is supplied by the signal transmitted by the reader.
- Semi-active – a power supply is present to power the microcontroller on the tag. The power required to transmit the signal is still provided by the reader's signal. This provides an intermediate option between passive and active modes, permitting additional computational processing but still restricting the distance required between the reader and tag.
- Active – a power supply is present to power both the microcontroller and antenna. This enables both improved computational processing and distance between the reader and the tag. The presence of the power supply, however, limits the operational lifetime of the tag. With no power supply, passive tags have a significantly longer lifetime than their active counterparts.

The passive tags function by using a property called inductive coupling. The signal being transmitted by the RFID reader induces a current in the antenna of the RFID tag. This current is then used to charge an on-board capacitor that subsequently provides the necessary voltage to power the microcontroller and transmit the resulting information back to the reader. An illustration of the process is presented in Fig. 5.14. Obviously, given the very low power, the operational distance between the reader and tag is merely a couple of centimetres. Indeed, practical implementations of this system, such as the London Underground Oyster card, recommend physical connection with the tag in order to ensure the RFID tag is brought in close enough range for the reader to successfully complete its task.

Whilst passive tags are excellent from a cost and lifetime perspective, the lack of significant power, both in terms of providing sufficient computational processing and the distance over which they can operate, means they are unsuitable from an authentication perspective. Whilst initially the computational processing required to perform the necessary authentication between the tag and reader would have been the most significant restriction, researchers have focused significant effort on

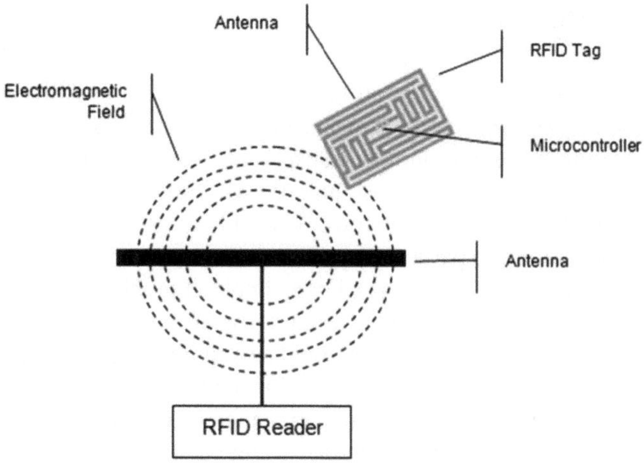

Fig. 5.14 Operation of an RFID token

developing low-cost (in terms of processing) security mechanisms to provide the necessary protection against attacks (Alomair et al. 2011). At present, the issue of utilising passive tokens for transparent authentication resides with the distance required between the tag and reader. Whilst some implementation scenarios could be developed – for instance, implementation of the tag within a watch, the reader would be contained within the mobile device – there is no guarantee a person would use the same hand that they are wearing the watch on for taking a call. Active tags could be implemented over a wider variety of options but can still restrict the operational distance as required – you wouldn't want the operational distance to be in metres, otherwise possible misuse could result.

Corner and Noble published a paper in 2005 entitled 'Protecting File Systems with Transient Authentication' that proposes the concept of utilising a wearable RFID token for use in continuous authentication of the user on a laptop or other computing device. The paper and others that have followed, such as Sun et al. (2008) and Guo et al. (2011), focus upon the computational overhead of the proposed protocols used to provide secure authentication. Suffice to say, such mechanisms are possible.

The use of RFID as an authentication technology does introduce additional hardware requirements that much of today's technology base is lacking. This subsequently has obvious implications over the cost of such a technology. Whilst RFID is certainly not ubiquitous at the time of writing, it is a technology being increasingly incorporated in a range of devices. Nokia released their first mobile phone with integrated RFID in 2004 (Nokia 2004). In 2007, Samsung reported to be developing RFID for mobile devices and more recently Apple has filled a patent regarding the use of RFID with the iPhone (Nystedt 2007; Evans 2010; Rosenblatt and Hotelling 2009). As mobile phones are amongst the most ubiquitous technology platforms, its inclusion is highly suggestive of wide adoption.

5.9 Other Approaches

The techniques described thus far all have direct application within a transparent authentication mechanism. It is not meant to be a definitive set of techniques but rather an in-depth discussion on how transparency can be applied. Indeed, as the field of authentication evolves, new techniques will be devised, many of which are likely to have modes of operation that would enable transparency.

The implementation of a particular approach has a significant effect over its applicability in this context of transparency. For example, if mobile phone manufacturers decided to include only a rear-facing camera, facial recognition would not be a usable, transparent approach – as a user would be required to turn the handset around. In this same manner, whilst some authentication approaches are not currently viable for transparent authentication, future implementations could well change this. For instance, fingerprint recognition is an intrusive approach that requires a user to physically place their finger onto the capture device. Implementations to date have always included the sensor in such a place on the mobile phone that requires a user to physically replace his finger/hand in order to provide the sample. If the sensor were replaced, perhaps in a location on the handset where a user's finger would more naturally reside, the finger could reside on the sensor whilst being held, and thus provide a mechanism for capturing fingerprint samples without requiring the user to intrusively provide the sample. A company, AuthenTec, has developed a technology referred to as Smart Sensors that integrates fingerprint recognition with additional functionality (AuthenTec 2011). For instance, each finger is associated to a different application, upon presenting the appropriate finger; an application will start and log the user in automatically. Whilst not completely transparent, as the user has to intrusively do something, it can be argued that the action being performed by the user is starting the application – an indirect consequence being the secure authentication of the user using fingerprint recognition. This more intelligent integration of functionality provides the user with a purpose to authenticate because it is intertwined with the operation they want to undertake, rather than merely authenticating for the sake of it and then having to select and start the application. A similar approach could also be applied through integrating fingerprint sensors with laptop mouse pads, generating samples as the user moves and clicks the mouse.

Ear geometry is another approach that with careful sample selection could provide a transparent approach. As highlighted in Sect. 4.4.3, ear geometry works much in the same way as facial recognition. By taking an image of the ear, particular features of the ear are then identified and extracted. Whilst under normal circumstances obtaining an image of the ear would require the user to take a picture of the ear – a challenging task given they would not be able to see what they are doing. However, particular interactions with a device could lead to such a sample being captured. Conceptually, the most applicable scenario would be with regard to mobile phones and telephone calls. The process of making or receiving a call would involve a person having to lift the handset to the ear. During that process, the front-facing camera has the opportunity to capture a sample. Figure 5.15 illustrates a series of images captured from the point of calling a telephone and the call connecting.

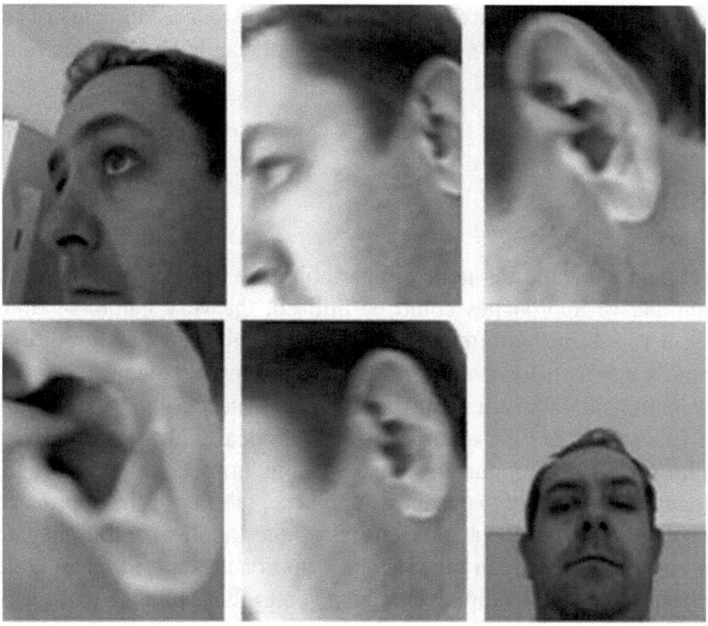

Fig. 5.15 Samples created for ear geometry

Obviously, given such a series of images presents a more challenging task for the ear geometry technique. In the first instance, the technique would need to identify which of the images was the most appropriate to utilise by applying feature identification and extraction algorithms. As the environment within which the sample is being taken is more dynamic, additional measures would be required to allow for rotation, normalisation and quality estimation of the sample to enable a consistent and subsequently more reliable feature vector to be obtained. However, these aspects are not insurmountable – indeed many biometric techniques include these.

Gait recognition is also another possibility for transparent authentication, but not in its usual format, as this requires a camera imaging a person walking at a distance. However, research studies have also been undertaken that perform gait recognition using the sensors attached to the body. Whilst in initial studies this involved the use of additional hardware strapped onto the body itself more recent research has utilised the built-in sensors of mobile phones to perform the task. A study in 2010 utilised the Google G1 phone, which has an accelerometer for measuring acceleration in three axes (Derawi et al. 2010). The study involved 51 participants walking up and down a 37 m corridor over two sessions carrying the mobile phone on their hip. They reported an EER of 20%. Whilst not the most impressive EER, this is a preliminary study using off-the-shelf technology (which few previous studies have achieved to date) and will certainly improve in the future. Gafurov et al. (2006) has published a study achieving an EER of 5%, using an ankle-based sensor. Whilst this might appear inconvenient, obtrusive and costly in the first instance, companies such as Nike are incorporating sensors within their footwear as a means of measuring distance run etc. (Nike sensor). These same sensors could also include functionality for enabling gait recognition.

5.10 Summary

Transparent authentication is not a simple objective to achieve. Whilst fundamentally, a number of approaches could be utilised, the very nature of transparency introduces a variability within the process that biometrics have traditionally performed poorly against. However, as research improves and the need exists to deploy biometric approaches across a variety of application scenarios, vendors are developing more robust and flexible algorithms that can deal with such variability within systems. Improvements in image rotation, scaling, normalisation and segmentation all enable improvements in the usability and acceptability of the approaches.

It is also evident from the analysis of techniques applicable to transparency that behavioural-based rather than physiological approaches are more suitable in general. Furthermore, consideration has to be given to the availability of the necessary hardware. Techniques that utilise pre-existing hardware have an obvious advantage and cost benefit. As highlighted in Table 5.19, techniques that are appropriate for transparency tend to belong to the category of techniques whose performance is poorer when compared to others.

However, transparency is not just about the individual techniques that enable transparency. It also requires careful consideration of its context. It requires a wider and higher-level perspective of the authentication requirements in order to ensure appropriate use of the technology. For instance, it requires a situational awareness so that relevant capture technologies can be enabled at the required time. It also needs to understand how to use those biometric inputs to get the most appropriate understanding of a user's legitimacy within the context in which they are using the device.

As such, transparency is more than simply an authentication technique, but rather a framework under which applicable authentication techniques would reside.

Table 5.19 Transparency of authentication approaches

Technique	Sample acquisition	Performance	Transparent
Behavioural profiling	✓	Low	✓
Ear geometry	✓	High	×
Facial recognition	✓	High	✓
Fingerprint recognition	×	Very high	×
Gait recognition	✓	Low	✓
Handwriting recognition	✓	Medium	✓
Iris recognition	×	Very high	×
Keystroke analysis	✓	Low	✓
Retina recognition	×	Very high	×
Speaker recognition	✓	High	✓

The framework has the capacity to understand its context and apply appropriate mechanisms both in the capture and response. In this manner, it is necessary to utilise a variety of authentication techniques. On occasions this might mean simultaneously combining appropriate techniques to improve overall performance, or others simply as a function of the interaction that is being performed by the user, with the outcome of one technique being used to decide whether further verification is required. This combinational use of biometrics is referred to as multibiometrics and is the subject of the following chapter.

References

Akkermans, A., Kevenaar, T., Schobben, D.: Acoustic ear recognition for person identification. In: Fourth IEEE Workshop on Automatic Identification Advanced Technologies (AutoID'05), Buffalo (2005)

Alomair, B., Lazos, L., Poovendran, R.: Securing low-cost RFID systems: an unconditionally secure approach. Journal of Computer Security – special edition on RFID systems, **19**(2), 229–257 (2011)

Aupy, A., Clarke, N.L.: User authentication by service utilisation profiling. In: Proceedings of ISOneWorld 2005, Las Vegas (2005)

AuthenTec: Smart sensors. AuthenTec. Available at: http://www.authentec.com/Products/SmartSensors.aspx(2011). Accessed 10 Apr 2011

Bishop, M.: Pattern Classification and Machine Learning. Springer, New York (2007)

Bolton, R., Hand, D.: Statistical fraud detection: a review. Stat. Sci. **17**(3), 235–255 (2002)

Brown, M., Rogers, J.: User identification via keystroke characteristics of typed names using neural networks. Int. J. Man Mach. Stud. **39**(6), 999 (1993)

Chan, P., Fan, W., Prodromidis, A., Stolfo, S.: Distributed data mining in credit card fraud detection. IEEE Intell. Syst. **14**(6), 67 (1999). Special Issue on Data Mining

Cho, S., Han, C., Han, D., Kin, H.: Web based keystroke dynamics. Identity verification using neural networks. J. Organ. Comput. Electron. Commer. **10**(4), 295–307 (2000)

Clarke, N.L., Furnell, S.M.: Keystroke dynamics on a mobile handset: a feasibility study. Inform. Manage. Comput. Secur. **11**(4), 161–166 (2003)

Clarke, N.L., Furnell, S.M.: Authenticating mobile phone users using keystroke analysis. Int. J. Inf. Secur. **6**(1), 1–14 (2006)

Clarke, N.L., Mekala, A.R.: The application of signature recognition to transparent handwriting verification for mobile devices. Inform. Manage. Comput. Secur. **15**(3), 214–225 (2007)

Clarke, N.L., Karatzouni, S., Furnell, S.M.: Transparent facial recognition for mobile devices. In: Proceedings of the 7th Security Conference, Las Vegas (2008)

Corner, M., Noble, B.: Protecting file systems with transient authentication. J. Wirel. Netw. **11**(1–2), 7–19 (2005)

Derawi, M., Nickel, C., Bours, P., Busch, C.: Unobtrusive user-authentication on mobile phones using biometric gait recognition. In: Sixth International Conference on Intelligent Information Hiding and Multimedia Signal Processing. IEEE, Darmstadt, Germany, 15–17 Oct 2010. Available at: http://ieeexplore.ieee.org/xpl/freeabs_all.jsp?arnumber=5638036 (2010). Accessed 10 Apr 2011

Distimo.: Our presentation from mobile world congress 2010 – mobile application stores state of play. Available at: http://blog.distimo.com/2010_02_our-presentation-from-mobile-world-congres-2010-mobile-application-stores-state-of-play/ (2010). Accessed 10 Apr 2011

Doddington, G.: Speaker recognition based on idiolectal differences between speakers. In: Proceedings of Interspeech, vol. 4, Aalborg, (2001)

Dowland, P.: User authentication and supervision in networked systems. PhD thesis, University of Plymouth, Plymouth, 2004

Dowland, P., Furnell, S.M.: A long-term trial of keystroke profiling using digraph, trigraph and keyword latencies. In: Proceedings of the 19th International Conference on Information Security (IFIP SEC), Toulouse, France (2004)

Eagle, N., Pentland, A., Lazer, D.: Inferring social network structure using mobile phone data. Proc. Natl. Acad. Sci. **106**, 15274–15278 (2009)

Evans, J.: Yet another RFID patent hints Apple has an iPhone wallet plan. 925 LLC. Available at: http://www.9to5mac.com/29306/yet-another-rfid-patent-hints-apple-has-an-iphone-iwallet-plan/ (2010). Accessed 10 Apr 2011

Gafurov, D., Helkala, K., Sondrol, T.: Gait recognition using acceleration from MEMS. In: The First International Conference on Availability, Reliability and Security (ARES 2006). Available at: http://ieeexplore.ieee.org/xpl/freeabs_all.jsp?arnumber=1625340 (2006). Accessed 10 Apr 2011

GB 2375205: Determining identity of a user. UK Patent Office. Available at: http://v3.espacenet.com/publicationDetails/biblio?DB=EPODOC&adjacent=true&locale=en_gb&FT=D&date=20021114&CC=WO&NR=02091310A1&KC=A1 (2002). Accessed 10 Apr 2011

Gosset, P. (ed.): ASPeCT: Fraud Detection Concepts: Final Report. Doc Ref. AC095/VOD/W22/DS/P/18/1 (Jan 1998)

Gross, R.: Face databases. In: Li, Stan Z., Jain, Anil K. (eds.) Handbook of Face Recognition. Springer, New York (2005)

Guo, M., Liaw, H., Deng, D., Chao, H.: An RFID secure authentication mechanism in WLAN. Comput. Commun. **34**, 236–240 (2011)

Jain, A., Patrick, F., Arun, R.: Handbook of Biometrics. Springer, New York (2008). ISBN 978-0-387-71040-2

Joyce, R., Gupta, G.: Identity authentication based on keystroke latencies. Commun. ACM **39**, 168–176 (1990)

Karatzouni, S., Clarke, N.L., Furnell, S.M.: Keystroke analysis for thumb-based keyboards on mobile devices. In: Proceedings of the ISOneWorld Conference, Las Vegas (2007a)

Kou, Y., Lu, C., Sirwongwattana, S., Huang, Y.: Survey of fraud detection techniques. In: Proceedings of the 2004 IEEE International Conference on Networking, Sensing and Control, Taipei, Taiwan, pp. 749–754 (2004)

Leggett, G., Williams, J.: Verifying identity via keystroke characteristics. Int. J. Man-Mach. Stud. **28**(1), 67–76 (1988)

Li, F., Clarke, N.L., Papadaki, M., Dowland, P.: Behaviour profiling for transparent authentication on mobile devices. In: Proceedings of the 10th European Conference on Information Warfare (ECIW), Tallinn, Estonia, pp. 307–314, 7–8 July (2011)

Lu, X., Jain, A., Colbry, D.: Matching 2.5D face scans to 3D models. IEEE Trans. Pattern. Anal. **28**(1), 31–43 (2006)

Moreau Y., Verrelst, H., Vandewalle, J.: Detection of mobile phone fraud using supervised neural networks: a first prototype. In: International Conference on Artificial Neural Networks Proceedings (ICANN'97), Lausanne, Switzerland, pp. 1065–1070 (1997)

Napier, R., Laverty, W., Mahar, D., Henderson, R., Hiron, M., Wagner, M.: Keyboard user verification: toward an accurate, efficient and ecological valid algorithm. Int. J. Hum. Comput. Stud. **43**(2), 213–222 (1995)

NIST: Speaker recognition evaluation. NIST. Available at: http://www.itl.nist.gov/iad/mig/tests/spk/ (2011). Accessed 10 Apr 2011

Nokia: Nokia mobile RFID kit. Nokia Corporation. Available at: http://www.nokia.com/BaseProject/Sites/NOKIA_MAIN_18022/CDA/Categories/Business/DocumentCenter/_Content/_Static_Files/rfid_kit_one_pager_v_2_0.pdf (2004). Accessed 10 Apr 2011

Nystedt, D.: Samsung develops RFID chip for cell phones. PCWorld. Available at: http://www.pcworld.com/article/139972/samsung_develops_rfid_chip_for_cell_phones.html (2007). Accessed 10 Apr 2011

Obaidat, M.S., Sadoun, B.: Verification of computer user using keystroke dynamics. IEEE Trans. Syst. Man Cybern. B Cybern. **27**(2), 261–269 (1997)

Ord, T., Furnell, S.M.: User authentication for keypad-based devices using keystroke analysis. In: Proceedings of the Second International Network Conference (INC 2000), Plymouth (2000)

Provost, F., Aronis, J.: Scaling up inductive learning with massive parallelism. Mach. Learn. **23**, 33–46 (1996)

Reynolds, D., Quatieri, T., Dunn, R.: Speaker verification using adapted Gaussian mixture models. Digit. Signal Process. **10**, 19–41 (2000)

Rodwell, P., Furnell, S.M., Reynolds, P.: A non-intrusive biometric authentication mechanism utilising physiological characteristics of the human head. Comput. Secur. **26**(7-8), 468–478 (2007)

Rosa, L.: Biometric source code. Advanced source code. http://www.advancedsourcecode.com/ (2008). Accessed 10 Apr 2011

Rosenblatt, M., Hotelling, S.: Touch screen RFID tag reader. US Patent 2009167699. Available at: http://worldwide.espacenet.com/publicationDetails/originalDocument?FT=D&date=20090702&DB=EPODOC&locale=en_T1&CC=US&NR=2009167699A1&KC=A1 (2009). Accessed 10 Mar 2011

Saevanee, H., Bhattarakosol, P.: User authentication using combination of behavioural biometrics over the touchpad acting like touch screen of mobile device. In: Proceedings of the 2008 International Conference on Computer and Electrical Engineering, Washington (2008)

Shin, C., Yun, U., Kim, H., Park, S.: A hybrid approach of neural network and memory-based learning to data mining. IEEE Trans. Neural Netw. **11**(3), 637–646 (2000)

Spillane, R.: Keyboard apparatus for personal identification. IBM Tech. Discl. Bull. **17**, 3346 (1975)

Stolfo, S.J., Wei, F., Wenke, L., Prodromidis, A., Chan, P.K.: Cost-based modeling for fraud and intrusion detection: results from the JAM project. In: DARPA Information Survivability Conference and Exposition, 2000. DISCEX '00. Proceedings, vol. 2, Hilton Head, SC, pp. 130–144 (2000)

Stormann, C.: Fraud Management Tool: Evaluation Report. Advanced Security for Personal Communications (ASePECT), Deliverable. 13, Doc Ref. AC095/SAG/W22/DS/P/13/2 (1997)

Sun, D., Huai, J., Sun, J., Zhang, J., Feng, Z.: A new design of wearable token system for mobile device security. IEEE Trans. Consum. Electron. **54**(4), 1784–1789 (2008)

Turner, I.: Eye net watch launch PDALok for IPAQ. Computing.Net. Available at: http://www.computing.net/answers/pda/eye-net-watch-launch-pdalok-for-ipaq/208.html (2001). Accessed 10 Apr 2011

US 5,787,187. Systems and methods for biometric identification using acoustic properties of the ear canal. US Patent (1998)

Chapter 6
Multibiometrics

6.1 Introduction

The ability to capture biometric samples transparently is a function of what the user is doing and the hardware available to provide the capture. As Fig. 6.1 illustrates, within a mobile device, an opportunity exists to capture face, voice, typing and behavioural characteristics (and possibly handwriting recognition). Some of these samples can be captured during differing user interactions with the device. For some interactions, multiple samples for use by different biometric approaches can be captured, in others, multiple samples of the same biometric, or indeed just a single sample for use by a single biometric. For example, typing a text message would provide samples for keystroke analysis, behavioural profiling and facial recognition. Playing a game on the device would provide the opportunity to capture multiple images of the face. Finally, simply checking the device to read a quick email might only provide the chance to capture a single facial image.

These different scenarios provide the opportunity for the transparent authentication system (TAS) to capitalise upon the multiple samples to provide more robust confidence in the authentication decision. The use of multiple samples is broadly referred to as multibiometrics. Multibiometrics can be divided into six categories:

- Multi-modal
- Multi-sample
- Multi-algorithmic
- Multi-instance
- Multi-sensor
- Hybrid

The first and second approaches relate to the earlier example: the use of multiple biometric approaches is referred to as multi-modal, and the use of multiple samples of the same biometric modality is referred to as multi-sample. The third approach of multi-algorithmic refers to utilising more than one algorithm during the matching phase. This can permit the system to capitalise upon the strengths of particular

N. Clarke, *Transparent User Authentication: Biometrics, RFID and Behavioural Profiling*,
DOI 10.1007/978-0-85729-805-8_6, © Springer-Verlag London Limited 2011

Fig. 6.1 Transparent authentication on a mobile device

classification approaches and subsequently provide a more robust decision. For instance, some facial recognition algorithms have a better performance with images with more significant facial orientations or changes in illumination. Multi-algorithmic approaches do not have to rely on a single approach but can combine the unique properties of each.

Multi-instance approaches are similar to multi-sample approaches but refer to the use of multiple subtypes of the same biometric, for instance, the left and right iris or the index finger of the left and right hand. This approach is useful particularly in situations where the user population is particularly large. Rather than relying upon a single biometric instance to provide sufficient distinctiveness to discriminate the whole population, the use of multi-instances simplifies the problem. The Integrated Automated Fingerprint Identification System (IAFIS) used by the US FBI processes the fingerprints from all ten finger and thumb prints. The system has over 66 million subjects stored in its database (FBI 2011). Whilst the need by the FBI to store all ten prints is partly a legal obligation, the additional information provided to the biometric system also aids in the quick indexing and searching of profiles.

Multi-sensor approaches utilise more than one sensor to capture a single biometric trait. This approach is useful when the different sensors are able to bring complementary information to the classification problem. For instance, in fingerprint recognition, optical and capacitive sensors have been used with significant improvement in overall performance (Marcialis and Roli 2004).

Finally hybrid systems utilise a combination of the aforementioned approaches, for example, the combination of a multi-modal approach and multi-algorithmic approach. Each individual technique utilises multiple classification algorithms to optimise the individual response, but those responses are also combined with other biometric techniques. Hybrid systems, if designed correctly, represent the most robust and strongest (in terms of recognition performance) biometric systems. They do, however, also represent the most complex ones, requiring increased storage, processing and architectural complexity.

In the intrusive domain of biometrics, the capture of samples can be performed in a synchronous and asynchronous approach. The former refers to capturing the biometric samples simultaneously or in parallel. The latter approach refers to the user interacting with each biometric capture device sequentially. In reality, for a TAS, both approaches could effectively be happening at any point in time so it is difficult to classify the system in that fashion. There are also multiple approaches to actually using these samples. It is possible to combine the output of the matching subsystems from multiple algorithms or the result from multiple approaches after the decision subsystem. It is also possible to take the output of one decision from a single biometric sample to inform whether a subsequent sample needs to be analysed. For instance, should the matching result of the first biometric result in a very high confidence in the authenticity of the user, there could be no need for further verification. Should the confidence be lower, further verification would be useful. In essence, a TAS could be multi-modal, multi-sample, multi-algorithmic, multi-instance, hybrid or uni-modal (the standard single-sample single-biometric approach) at different points in time. The decision upon which to use largely relies upon whether sample data is available for the system to use (although other issues such as system complexity are also present – these will be considered in Part III). This in turn is dependent upon whether the user is interacting with the device.

As transparent authentication enables the capture of multiple samples as a natural function of its operation, it seems prudent to consider how multibiometrics can play an advantageous role within the system.

6.2 Multibiometric Approaches

The individual weaknesses of biometric approaches identified that no single biometric (or indeed authentication technique) could provide sufficient verification of a complete population. One of the key aims of transparent authentication is to provide a framework that enables a variety of approaches to be utilised in order to mitigate the problem. Within the biometric industry, the focus was placed upon multibiometrics – capitalising upon the strengths of two or more approaches to circumvent issues. Each multibiometric approach, multi-modal, multi-instance and multi-algorithmic, has its own set of advantages that help achieve a particular set of objectives.

For multi-modal approaches, the use of more than one biometric technique enables the system to reduce the problem of universality. All biometric techniques suffer from a (typically small) proportion of the population unable to present the particular biometric trait. Facial recognition is perhaps one of the few exceptions. The use of multi-modal approaches provides the ability of the system to rely upon more than one biometric technique. For instance, the International Civil Aviation Authority suggests the use of fingerprint and facial recognition technologies. Should users be unable to present reliable and consistent fingerprint samples, a secondary measure is present. The performance of the overall approach, for users who can provide both traits, can also be improved. The need to provide more than one sample will also help in preventing spoof attacks because an attacker needs to circumvent not just one technique but two or more. Whilst combining them does not increase the difficulty of circumventing the individual techniques, it does increase the workload and effort required from the attacker. After cost, which will be discussed later, the principal disadvantage with multi-modal approaches is acceptability. The inclusion of multiple sensors from different modalities within the normal intrusive application scenario places an additional burden upon the individual. Within a TAS, however, this is not the case.

However, whilst utilising transparent systems improves the level of convenience, it is traded off with variations in the samples collected. Within intrusive approaches, the system will ensure samples from all modalities are utilised. They can then be processed and a decision made on that specific set of modalities. Within a transparent system, it is not possible to guarantee that the required samples will all be present. Using the earlier example of composing text message, keystroke analysis, behavioural profiling and facial recognition samples can be used. However, should the user be texting someone at night whilst outside, there could be insufficient illumination to provide a facial sample. In this scenario, the system would only have two of the three samples required for authentication. Whilst this is not fundamentally an issue technically, given the number of possible usage scenarios and subsequent samples that can be captured, it does introduce a significant increase in the complexity of the system. How this can be combated will be discussed further in Chap. 8.

For multi-sample approaches, whilst the performance cannot be improved beyond the base performance of the technique, the approach is capable of providing stronger performance through the combination of matching scores from multiple samples of the same biometric. Within a TAS, a user could be undertaking a video conference call via Skype. During the call, the web camera is in a position to take a series of facial images. Whilst one approach could be to implement a uni-modal system that simply performs authentication after every sample is taken, another approach could be to capture a series of images and process them simultaneously. The latter approach has the advantage of smoothing out any irregularities that exist within the samples. For instance, in a uni-modal system, should an image result in a false rejection error being generated, the system would reduce the confidence level in the identity of the user potentially forcing the user to re-authenticate. By using a multi-sample approach, a more robust decision can be established on the identity of the user, thereby reducing errors. Multi-sample approaches are also useful for behavioural-based biometrics. These approaches tend to suffer from larger variations in input samples. The use of multiple samples provides a mechanism for reducing the impact any single (anomalous) sample might have.

Within a transparent system, it is highly likely that the multiple-sample scenario will be available. As the system is simply capturing samples during each and every user interaction, the biometric data collected is going to increase and incorporate a range of sample data from the limited number of sensors available. Whilst the transparent nature of the capture requires significant measures in place regarding the sample quality, a transparent system would be able to utilise a multi-sample approach. Much like the multi-modal approach, the extent to which it can use the multi-sample approach will depend upon the particular user interactions and the time that has elapsed.

For the same reasons as the multi-sample approach, multi-algorithmic approaches also provide a more robust result. But rather than focusing upon the potential weaknesses between samples, multi-algorithmic approaches look to improve upon the weaknesses present in a classification algorithm. The use of multi-algorithmic approaches can also be referred to as a multi-classifier system and such systems have been widely recognised as providing significantly better performance over their uni-algorithmic counterparts. The basis for this improvement is largely based upon utilising classification algorithms that base or weight their matching score on different attributes of the sample or feature vector. As highlighted in the previous chapter, each biometric technique has a variety of approaches to achieve classification. Each approach capitalises upon particular attributes of the sample. If algorithms, which focus upon different aspects are combined they can complement each other. For instance, speaker recognition systems have relied upon either spectral or higher-level classification approaches. Both approaches look at different attributes of the biometric sample. A spectral approach focuses upon the vocal characteristics of the sample, whilst the higher-level language approach tends to focus upon how and what is being said. A combination of both approaches would constructively support each other.

The use of multi-algorithmic approaches in a transparent system is no different to their application within an intrusive application scenario. They have minimal impact upon user convenience in terms of sample capture and provide a more robust measure of identity, thereby reducing inconvenience through false rejections. In a TAS, they could be implemented in one of two approaches:

1. A complete solution from a biometric vendor. The matching subsystem is inherently based upon a multi-classifier approach. The output from this particular biometric technique is a single value after the multi-algorithmic approach has been applied.
2. The TAS permits the use of multiple vendor-matching systems. The TAS would provide the sample to each and then collate the matching score.

The former is simpler to implement but restricts the system to a particular biometric vendor for that biometric trait. The latter enables the system to utilise as many vendor-matching subsystems as required, enabling a selection of the best-performing algorithms across vendors to be utilised. However, the latter approach also requires the TAS to take responsibility for combining the resultant matching scores from each algorithm. It is also the more expensive option, as each vendor algorithm would need to be purchased.

Given the increasing trend towards multi-algorithmic approaches, many biometric vendors are likely to contain multi-algorithmic approaches in order to improve performance.

Multi-instance-based approaches are less likely to appear within a TAS; however, given the transparent operation of the system, some of the methods proposed to handle the variability in data might give rise to a multi-instance approach. An example of this is the potential use of keystroke analysis. The technique can operate in a static or dynamic mode, the former being based upon a consistent string of characters such as a password and the latter upon a text-independent fashion. The performance of static-based approaches has outperformed those of dynamic-based approaches fundamentally because the variability in the users' data (the intra-class variation) is smaller (as one might naturally expect).

However, within a transparent system, the sample must be captured non-intrusively, so on the face of it, a static-based approach seems less feasible than the dynamic one. However, with sufficient input data, it is possible to design a pseudo-dynamic approach that is comprised of a number of static-based approaches. As sufficient repetitions of a single word are typed (as keystroke analysis typically requires far larger numbers of samples to create a template) a static-based template can be created for that word. As time goes on, a database of static-based profiles can be developed. Subsequent pseudo-dynamic authentication can then be performed, as long as one of the words for which a template exists appears in the text.

In situations where more than one word appears within the text, for which a template exists, a multi-instance approach can be implemented. If the same word appeared twice in the text and these samples were used, it would be classed as a multi-sample rather than multi-instance approach. In situations, such as a user typing a report on a computer, it is highly likely for profiled words to appear frequently. Similar to a multi-sample approach, rather than relying upon a single sample, for which with behavioural biometrics suffer from increased sample variability, this approach enables a smoothing out of individual irregularities.

From an acceptance, robustness and performance perspective, a TAS system would benefit most from utilising the hybrid mode where a variety of these multibiometric modes could be implemented to suit particular operational characteristics. This will have an obvious effect upon the system complexity, as with more modes of operation and additional classification algorithms come increased architectural complexities in order for them to work together within a single system. Additionally, the approach places a burden on the system responsible for processing these data in terms of the computational requirements. Whilst desktop-computing systems could arguably cope with such systems, could the more computationally restricted mobile devices? A thorough consideration of all these operational parameters will be presented in Part III.

From a processing perspective it is possible to devise mechanisms that reduce the overall burden. Whilst sample acquisition in intrusive systems can be achieved in a synchronous and asynchronous approach, this same methodology can also be applied to the processing of samples. For instance, whilst capture might be performed in a synchronous mode (i.e. at the same time), the processing of the samples can be performed in an asynchronous or sequential mode. Whilst the mode of sampling has little effect upon a TAS – as samples are captured continuously in the background – the mode of processing could have a significant impact. In a cascade mode, the processing of samples takes place sequentially, as illustrated in Fig. 6.2. If given sample A, the authenticity of the user cannot be verified, the system can move to sample B, sample C and so on. At any point during the process, should the verification come back successfully, the process can stop, removing the need to further process samples. If the order of samples presented to the system is based

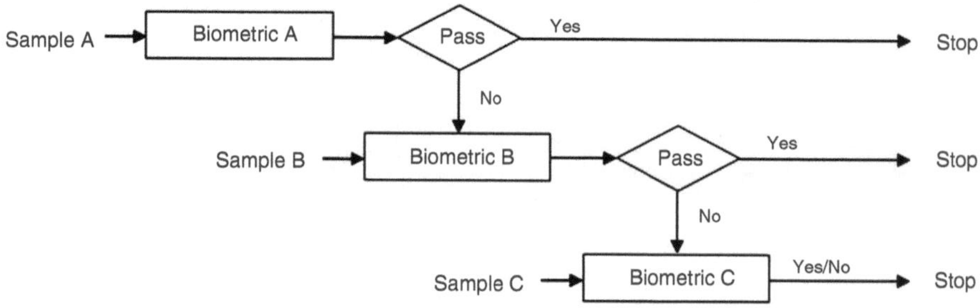

Fig. 6.2 Cascade mode of processing of biometric samples

upon the overall performance/confidence of the biometric technique (i.e. facial recognition outperforms behavioural profiling in terms of performance), the level of processing could be reduced.

6.3 Fusion

In each of the multibiometric approaches a requirement exists to combine some data, for instance, the matching scores from multiple algorithms or the decisions from multiple biometric approaches. The term fusion is given to this combination or fusing of data in order to enhance the overall performance of biometrics. The nature of the data fusion can occur effectively at any point within the biometric system:

- Sensor image
- Feature extraction
- Matching subsystem
- Decision subsystem

The type of fusion that is required will depend upon a variety of factors. However, certain types of multibiometrics can only implement fusion later on in the biometric process. For example, a multi-algorithmic approach requires data fusion at either the matching or decision subsystems. It would not be viable to fuse the data any earlier as it would not be possible to process the data through the different algorithms. Multi-sample approaches could utilise fusion at the sensor or feature extraction processes to produce a more robust image or feature vector for use by the matching subsystem. Figure 6.3 illustrates one approach to achieving matching score-level fusion – where the outputs from multiple classifiers are combined to provide a single result that is presented to the decision subsystem.

Sensor-level fusion is useful in certain applications for establishing or developing a more effective sample, such as the creation of a 3D facial image from two 2D images. Sensor-level fusion has limited use within a TAS due to the lack of multiple sensors that capture the same biometric trait.

For approaches with multiple samples of the same biometric trait, fusion at the feature extraction process can enable a more robust feature vector to be developed

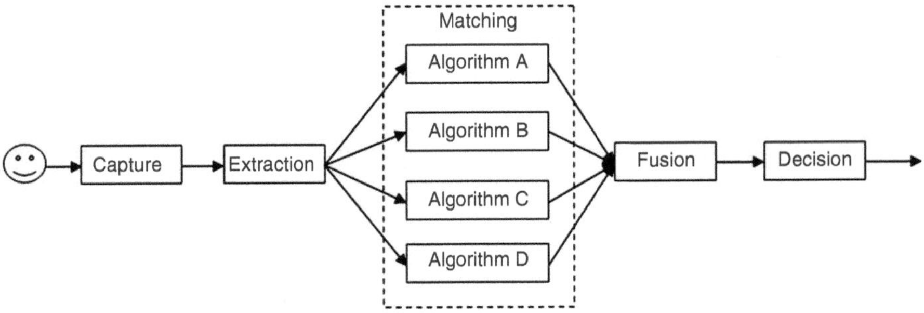

Fig. 6.3 Matching score-level fusion

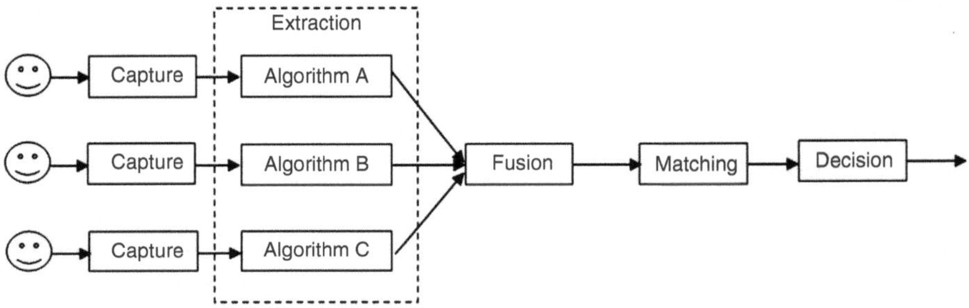

Fig. 6.4 Feature-level fusion

based upon a weighted average of the feature vectors generated individually (or by any other combinational algorithm they wish). Rather than putting each sample through the complete biometric process and obtaining a decision, feature-level fusion allows the creation of a single feature vector from a series of samples, as illustrated in Fig. 6.4. The biometric process then only needs to be performed once – reducing computational overheads (assuming the fusion process is less computationally intensive than several runs of the matching subsystem process – which invariably it is).

Feature-level fusion can also be applied to different modalities – effectively concatenating or appending them together. Some biometric vendors may create matching subsystems based upon a variety of features derived from different modalities – rather than using separate matching subsystems. In such an approach, the feature vectors of the two independent approaches need to be combined in order to be inputted into the matching subsystem. Such systems typically also include measures for reducing the feature vector to the most efficient feature size for the combined modalities, in order to achieve its task. The larger the feature vector, the more complex the resulting matching subsystem needs to be – a problem known as the curse of dimensionality (Bishop 2007). Therefore, it is necessary to ensure the feature vector is as efficient as possible to achieve the classification problem.

Matching- or score-level fusion takes the output of the resulting matching subsystems and combines the results prior to presenting to the decision subsystem. Of all the fusion approaches, match-level fusion is the most widely used. The approach enables multi-modal authentication with each modality being classified by a dedicated matching subsystem designed specifically for that approach. The most challenging problem with score-level fusion is how to interpret the outputs. If all the outputs from the various classifiers are identical, the combination of those results is relatively simple. However, invariably the use of different classifiers results in a different output being produced. For instance:

- The output from one classifier might be a measure of the similarity between two samples; for instance, a statistical-based classifier that provides a probabilistic value of how similar the two samples are. The larger the value, the more similar the two samples. Another classifier, based upon minimal distance, would output a distance measure. The smaller the distance, the more similar the samples.
- The output from another classifier might be a measure of their dissimilarity. Conversely to levels of similarity, these classifiers are designed to output the opposite of similarity measures.
- The range of values available to a classifier might vary considerably. For instance, the output of two popular neural network classifiers, the Feed-Forward Multi-Layered Perceptron (FF MLP) and Radial Basis Function (RBF) networks have an output that varies from 0 to 1 and 0 to ∞.

There are various techniques to solving this problem (Ross et al. 2006). A common approach is to develop some mechanism for normalising the outputs so that they can be combined in a consistent manner. One such technique is the min–max approach, where each output is scaled based upon the minimum and maximum value within its own output range. The resulting values are therefore proportional to the range of values that its own particular classifier is able to output.

For a TAS, match-level fusion offers a wide range of opportunities given the nature of the sample data that can be collected. The issue that arises is how the fusion process operates if all samples are not present. Whilst multi-algorithmic approaches will not be affected, multi-modal, multi-sample and multi-instance approaches will. How effective can the fusion process be when it is missing values at several of the outputs of the classifiers? The solution to the problem conceptually is one of two approaches:

1. To create a multi-modal/multi-sample/multi-instance system for every permutation or combination of inputs. For instance, if a system has the ability to capture three biometric traits, the TAS would create seven biometric systems or processes:

   (a) 3 uni-modal systems based upon each of the individual traits
   (b) A multi-modal system based upon traits 1 and 2
   (c) A multi-modal system based upon traits 1 and 3
   (d) A multi-modal system based upon traits 2 and 3
   (e) A multi-modal system based upon traits 1, 2 and 3

   In this approach, the TAS is able to cope with any combination of input sample, merely through the selection of the most appropriate system. However, as the number of approaches (i.e. multi-modal, multi-sample and multi-instance) increases, so do the number of biometric systems and the complexity.

2. Develop a fusion process that is able to cope with missing data. Neural network approaches can be utilised to model the output data from the matching subsystems. However, performance relies upon having sufficient training data to let the neural network learn how to behave. Having access to reliable training data that is truly representative of the problem the neural network is trying to solve is not always simple.

Decision-level fusion occurs at the end of the biometric process when each individual biometric system has provided an independent decision. Within a verification problem, this result is effectively a Boolean value. For that reason, decision-level fusion lacks the richness of information that match-level fusion has with raw scores. However, with some implementations, decision-level fusion is the only option. For example, in a multi-modal system, a system designer might choose to select a series of vendor-specific algorithms, the output from which is a decision rather than a match score.

A common technique used for fusion at the decision level is majority voting – each classifier has an equal weighting and the majority that agree with the decision win. A slight modification of that approach is the weighted majority voting – this approach takes into consideration the relative recognition performance of the underlying approaches. For example, a facial recognition approach performs better than a behavioural profiling approach and should therefore not have an equal weighting in the decision process. This weighted approach becomes particularly relevant in a TAS given the use of poorer performing behavioural approaches.

A TAS could therefore benefit from a variety of these approaches being utilised. The multi-sample approach could utilise fusion at the feature, match or decision level or a different model for each biometric modality. Multi-algorithmic approaches would usefully utilise match-level fusion in order to benefit from the additional information stored in the result over their decision-level fusion counterparts. Figure 6.5 illustrates one of many possible models a TAS could implement. What the combination of approaches utilised is in terms of modes of multibiometrics or methods of fusion is actually dependent upon what recognition performance is to be achieved. The important feature TAS must provide is a flexible framework that can support such variability on an individual basis.

6.4 Performance of Multi-modal Systems

Research into multi-modal systems has increased considerably over the last decade due to the enhancements in performance that can be experienced. Literature is available that directly compares the performance of multi-modal systems over their unimodal counterparts. To highlight these aspects, this section will describe the findings from three multi-modal studies involving

1. Finger and face modalities
2. Finger, face and hand modalities
3. Face and ear modalities

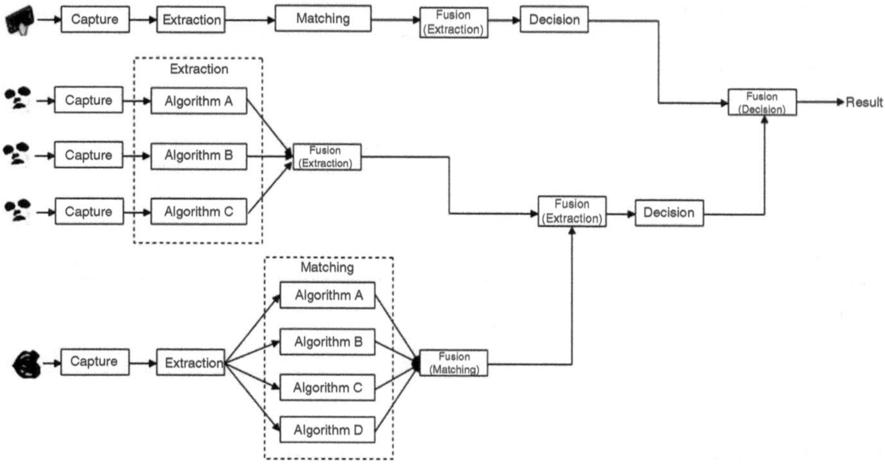

Fig. 6.5 A hybrid model involving various fusion approaches

Table 6.1 Multi-modal performance: finger and face

Classifier	FRR		EER (%)
	FAR = 1%	FAR = 0.1%	
Fingerprint	25	32	2.16
Face	59	100	3.76
Multi-modal	9	21	0.94

The first study is interesting because it involves a large population of users numbering almost 1,000 (Snelick et al. 2005). Few biometric studies tend to include such high levels of participation – particularly for multi-modal applications. Utilising a dataset with such a large number of participants provides for a more statistically reliable set of measurements and results. The application sourced datasets from two independent sources. The facial data were based upon the FERET database (a publicly accessible dataset used for testing facial recognition algorithms) (FERET 2011) and the fingerprint data came from a proprietary source. Whilst the two sets of data do not come from the same individual, their independent nature means that they can be combined (i.e. an individual's face has no impact upon the resulting fingerprint image). A standard methodology was utilised in the study in terms of calculating the performance characteristics – one user acted as the authorised user and the remaining as impostors. All users were given the opportunity of playing the authorised user. The match-level fusion approach was applied using a simple sum function (applied after normalisation). As illustrated in Table 6.1, the performance of the multi-modal approach is better than either of the uni-modal approaches.

The second study by Jain et al. (2005) studied the performance of a multi-modal system incorporating the finger, face and hand modalities. They utilised the Michigan State University (MSU) multi-modal database with a set of 100 virtual participants with finger, face and hand geometry samples. The study sought to evaluate the performance of the multi-modal approach against their uni-modal counterparts but also

Table 6.2 Multi-modal performance: finger, face and hand modalities

Classifier	FRR at a FAR = 0.1%
Finger	16.4
Face	32.3
Hand	53.2
Multi-modal (min–max norm)	2.2
Multi-modal (Tanh norm)	1.5

Table 6.3 Multi-modal performance: face and ear modalities

Classifier	TPIR (%)
Face	70.5
Ear	71.6
Multi-modal	90.9

to compare different normalisation strategies. The key results from the research are presented in Table 6.2. The fusion technique applied to the results was the sum of scores method. It can be seen from the table that the normalisation approach does indeed have a significant role to play in the performance of the multi-modal approach. Interestingly, all seven normalisation approaches tested in the study showed better levels of performance than the uni-modal systems. (The 'Tanh' function normalises the output on a non-linear scale from −1 to +1.)

The third study utilised face and ear modalities (Chang et al. 2003). This study is interesting for two reasons. It is one of the few studies that include the ear modality within the system. It also applies fusion at the sensor rather than the match level. The face and ear raw samples are concatenated together and processed by the feature extraction algorithm. An experiment based upon 197 participants found the multi-modal approach significantly outperformed the uni-modal systems. The methodology is based upon an identification rather than verification system, so the results are represented in terms of the true positive identification rate (TPIR) based upon being returned in the rank 1 position (Table 6.3).

Whilst studies are somewhat limited in multi-modal biometrics when compared to uni-modal biometrics, there is clear evidence that the performance of multi-modal systems significantly outperforms the standard approaches.

6.5 Summary

Multibiometrics offers an opportunity to overcome many of the weaknesses of individual uni-modal biometrics. They offer a variety of advantages including:

- Resolving issues with users who do not have or cannot present particular biometric traits.
- The ability to make authentication decisions on a wider set of information, including multiple samples of the same biometric, multiple instances of subtypes of the same biometric, multiple algorithms or multiple biometrics.

- Synchronous and asynchronous possessing modes, minimising the computational overhead and improving user acceptance through a reduction of user inconvenience.
- Providing a significant enhancement in the overall performance of the biometric system.

For use in a TAS, multibiometrics has an even greater role to play. Whilst multi-sample and multi-algorithmic approaches have universal advantages to both intrusive and transparent systems, intrusive systems suffer when looking to utilise multi-modal approaches due to the user inconvenience factor. In transparent systems, these samples are captured automatically and therefore provide a simple means of enabling multi-modal scenarios.

The implementation of multi-modal systems, however, is not a simple task within a TAS. Several aspects and considerations need to be taken into account. From an interoperability perspective, the TAS will need to communicate with the relevant biometric components provided by differing vendors within the system. For example, multi-algorithmic approaches will require the TAS to provide the appropriate biometric feature vector to each of the vendor-specific algorithms. From a wider perspective, multibiometric systems introduce a variety of architectural and operational considerations including how to manage a system using a variety of multi-modal, multi-algorithmic, multi-instance and multi-sample approaches and what computational overhead such a system is going to require, notwithstanding issues pertaining to privacy concerns and legalisation it will need to abide by.

In order to provide a fuller discussion of these issues, the following chapter will introduce the work being undertaken in the standardisation of biometrics, enabling vendor-independent multibiometric systems such as a TAS to exist. Part III of this text will focus upon the theoretical and practical aspects of implementing a TAS.

References

Bishop, C.M.: Pattern Classification and Machine Learning. Springer, New York (2007)

Chang, K., Bowyer, K., Sarkar, S., Victor, B.: Comparison and combination of ear and face images in appearance-based biometrics. IEEE Trans. Pattern Anal. Mach. Intell. **25**(9), 1160–1165 (2003)

FBI: Integrated automated fingerprint identification system (IAFIS). Federal Bureau of Investigation. Available at: http://www.fbi.gov/about-us/cjis/fingerprints_biometrics/iafis/iafis (2011). Accessed 10 Apr 2011

FERET: The facial recognition technology (FERET) Database. NIST. Available at: http://www.itl.nist.gov/iad/humanid/feret/feret_master.html (2011). Accessed 10 Apr 2011

Jain, A., Nandakumar, K., Ross, A.: Score normalisation in multimodal biometric systems. Pattern Recognit. **38**(12), 2270–2285 (2005)

Marcialis, G., Roli, F.: Fingerprint verification by fusion of optical and capacitive sensors. Pattern Recognit. Lett. **25**(11), 1315–1322 (2004)

Ross, A., Nandakumar, K., Jain, A.: Handbook of Multibiometrics. Springer, New York (2006)

Snelick, R., Uludag, U., Mink, A., Indovina, M., Jain, A.: Large scale evaluation of multimodal biometric authentication using state-of-art systems. IEEE Trans. Pattern Anal. Mach. Intell. **27**(3), 450–455 (2005)

Chapter 7
Biometric Standards

7.1 Introduction

The preceding chapters have identified the requirement for a more sophisticated method of authentication utilising a variety of authentication techniques rather than relying upon one. The advantages of doing so result in improvements in usability, acceptability and security; however, for such an approach to be feasible in practice, authentication techniques must operate in a truly interoperable fashion, where techniques have effectively 'plug "n" play' functionality.

Historically, the commercialisation of biometrics involved the creation of bespoke implementations that were carefully crafted into enterprise management systems. Indeed, this is still a large aspect of a vendor's product offering today. These systems were typically connected to a particular vendor's products, which not only ensured that all initial hardware and software systems were purchased from them but also linked the organisation to that specific vendor for an extended period of time. It was challenging for an organisation to buy a hardware-capture device from one company and integrate it into the extraction and classification algorithms of another company. The concept of off-the-shelf biometric solutions simply did not exist.

Whilst this situation has not wholly changed, in recent years a concerted effort has been placed upon standardisation, decoupling complete biometric systems into their consistent components and providing mechanisms for each to communicate and operate through a common format. Through this evolution and maturing of the biometric market, approaches such as multibiometrics can truly be deployed in a cost-effective manner.

7.2 Overview of Standardisation

It is clear that standardisation is a technology enabler and provides a mechanism for organisations to ensure they utilise the best configuration of components and future-proof the technology. The development of standards can be driven from a variety of

N. Clarke, *Transparent User Authentication: Biometrics, RFID and Behavioural Profiling*,
DOI 10.1007/978-0-85729-805-8_7,

Table 7.1 ISO/IEC JTC1 SC37 working groups

Group	Focus area
ISO/IEC JTC 1 SC 37 WG 1	Harmonised biometric vocabulary
ISO/IEC JTC 1 SC 37 WG 2	Biometric technical interfaces
ISO/IEC JTC 1 SC 37 WG 3	Biometric data interchange formats
ISO/IEC JTC 1 SC 37 WG 4	Biometric functional architecture and related profiles
ISO/IEC JTC 1 SC 37 WG 5	Biometric testing and reporting
ISO/IEC JTC 1 SC 37 WG 6	Cross-jurisdictional and societal aspects

directions including formal and informal Standards Development Organisations (SDOs) within both a national and an international context, industry and consumers. From an international perspective, the International Standards Organisation (ISO) and International Electrotechnical Commission (IEC) play a significant role in biometric standards via a joint partnership, as they do with many areas of joint interest. On a national front, there are SDOs such as the American National Standards Institute (ANSI), the British Standards Institute (BSI) and the National Institute of Standards and Technology (NIST) to name but a few. Informal SDOs such as the Biometric Consortium, BioAPI Consortium and the World Wide Web Consortium (W3C) exist to serve their membership and typically provide very industry-specific guidelines. In a variety of consumer or industry-sector domains, organisations exist that develop standards to meet specific purposes. For instance, the International Civil Aviation Organisation (ICAO) mandates a standard regarding the use and storage of biometrics within international travel documents (ISO/IEC 7501-1 2008).

The primary coordination of biometric standards is performed by ISO/IEC under Joint Technical Committee (JTC) 1 Subcommittee (SC) 37. As illustrated in Table 7.1, SC37 is comprised of six working groups covering key aspects from harmonising biometric vocabulary and developing data interchange formats to cross-jurisdictional and societal aspects. It is worth highlighting that the efforts of SC37 encompass the complete range of biometric applications: verification, identification, watch list and forensics. Some aspects of the standards are less relevant in the specific context of a Transparent Authentication System (TAS), and as such will not be discussed.

Within each working group, standards are developed and published through international collaboration. ISO/IEC have developed an onion model to represent the relationships between standard efforts. As illustrated in Fig. 7.1, the model has a series of seven layers, each one encapsulating the previous one (ISO 19784-1 2006a).

The innermost layer contains the Data Interchange Formats. These standards pertain to the interoperability of data between systems using the same modality. This will enable a common approach to the structure of biometric data and represents the lowest level of interoperability. The Data Structure layer refers to standards that define the wrappers for data exchange between systems or components. In order to appreciate the difference between the two layers, an analogy to the IP protocol can be used. The IP protocol relates to the Data Structure standards and formally structuring the payload of the IP packet would involve the Data Interchange Format standards. The Data Security Standards layer provides protection for the

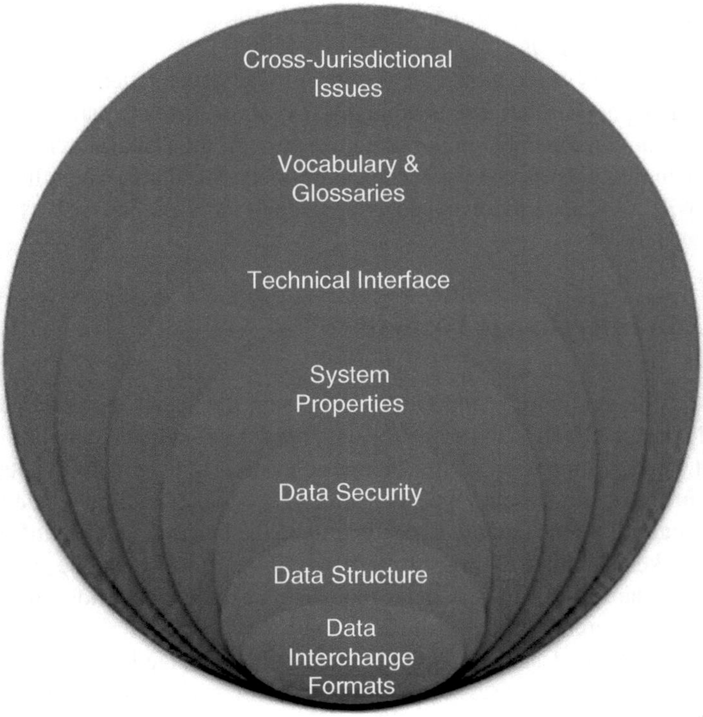

Fig. 7.1 ISO/IEC onion-model of data interchange formats

previous two layers in terms of the vulnerability of the biometric data. ISO/IEC JTC 1 SC 27 (IT Security Techniques) are responsible for defining these standards and they pertain to security countermeasures such as the appropriate use of cryptography to secure the communication channel. System properties standards assist in providing interoperability through the use of biometric profiles and evaluating the effectiveness of the approach within that context. The use of application-specific profiles permits more appropriate evaluations to be performed that enable uniform conformance testing. Without such profiles, comparing the performance of a biometric system, for example in airport identification, versus a system deployed for point-of-sale merchants would be unreliable. The Technical Interface Standards provide the mechanisms for the various biometric components to interoperate. The standards provide a framework for hardware and software components to communicate, with an ability to decouple the biometric system's sub-components from a single vendor. The penultimate layer refers to the standardisation of vocabulary and terms of reference to ensure all stakeholders have a uniform understanding of the terms. This avoids confusion and misunderstanding to develop over any aspect of the system and its performance. For instance, the false acceptance rate (FAR) and false rejection rate (FRR) are frequently misquoted as the match/classification performance in studies rather than the decision performance. The final layer in the model refers to all the higher-level non-technical aspects such as the impact of laws and treaties, privacy concerns relating to the handling and storage of biometric data, ethical issues and health and safety.

Whilst all layers of the model are pertinent to the development of biometric systems, including approaches discussed in this text, three layers in particular are key to establishing an interoperable multibiometrics system: the Data Interchange Format, Data Structure Standards and the Technical Interface Standards. The following three sections will focus upon these and describe them in more detail.

7.3 Data Interchange Formats

The core of the biometric standards focuses upon the actual representation of the biometric data itself. How to format the data for the individual techniques in such a way that other components of the biometric system will be able to interpret the information? The number and type of features required, even for a single biometric modality, can vary considerably depending upon the matching algorithm employed. Each bespoke matching algorithm often requiring its own unique combination. Without standardisation of this data, biometric capture, extraction and matching processes are inherently tied together.

The standard that deals with defining the data interchange format is ISO19794-1 (2006b), which is split into several parts for each biometric approach, including part 1 which describes the framework (as illustrated in Table 7.2). It may be noted by the publishing dates that many of these standards are relatively new. Indeed, the final few on the list are not current standards but work in progress, with the Committee Draft (CD) and Final Committee Draft (FCD) relating to particular stages in the creation and ratification of the standard. This highlights the significant degree of effort that has been undertaken over the past decade but also emphasises the maturity of the

Table 7.2 ISO/IEC biometric data interchange standards

ISO/IEC reference	Description
ISO/IEC 19794-1:2006	Part 1: framework
ISO/IEC 19794-2:2005/Amd 1:2010	Part 2: finger minutiae data
ISO/IEC 19794-3:2006	Part 3: finger pattern spectral data
ISO/IEC 19794-4: 2005	Part 4: finger image data
ISO/IEC 19794-5:2005	Part 5: face image data
ISO/IEC 19794-6:2005	Part 6: iris image data
ISO/IEC 19794-7: 2007	Part 7: signature/sign time series data
ISO/IEC 19794-8: 2006	Part 8: finger pattern skeletal data
ISO/IEC 19794-9: 2007	Part 9: vascular image data
ISO/IEC 19794-10: 2007	Part 10: hand geometry silhouette data
ISO/IEC FCD 19794-11	Part 11: signature/sign processed dynamic data
ISO/IEC NP 19794-12	Part 12: face identity (withdrawn)
ISO/IEC CD 19794-13	Part 13: voice data
ISO CD 19794-14	Part 14: DNA data

domain, with many biometric approaches still not defined – particularly the newer and less established behavioural biometrics.

The biometric process is somewhat more complicated than the stages of capture, extraction, matching and decision that are typically referred to, with many stages themselves comprising a number of processes. For instance, pre-processing of the sample data is a key step prior to extraction. Pre-processing allows several processes to take place including detection, alignment, segmentation and normalisation. However, where this should take place differs within implementations. For some, pre-processing and extraction of features can take place at the sensor, reducing the volume of data needing to be communicated. In others, to reduce sensor cost, the raw sample is communicated to a backend system for processing. A combination of both approaches is also possible, with pre-processing occurring at the sensor with the more computationally complex feature extraction taking place in the backend system. To allow for the varying forms in which the data could be present, the standard defines three processing levels of biometric data:

- Acquired data – data acquired from the sensor in its raw form.
- Intermediate data – processed raw data but not currently in the form required for matching. These data are further split into image or behavioural data. Image data pertain to biometric techniques that acquire a sample as an image. Information regarding the image format (e.g. JPEG, BMP, TIFF), image resolution, lighting conditions and distance from camera is included. With behavioural data, static images are replaced with dynamic actions such as dynamic signature recognition. Therefore templates are more attuned to time- or frequency-based data representation.
- Processed data – data that can be used by the matching subsystem directly. Also referred to as feature data.

The nature of each data interchange format is specific to the modality; however, a number of general considerations are relevant (ISO19794-1 2006b):

- Natural variability – no two samples are identical, so additional information such as tolerances could be useful to include within the data structure.
- Aging and usage duration – permanence of some biometric techniques is poor requiring updating of enrolment templates. A validation period is useful to ensure the sample is still appropriate to use.
- Enrolment conditions – important to define the minimal requirements such as resolution, number of minutiae or contrast.
- Feature extraction algorithms – if specified at this level (feature data), the methods for deriving these features need to be included within the data structure and be within acceptable tolerances.
- Feature-matching algorithms – if specified at this level, the methods for matching need to be defined to the extent necessary to enable interoperability.
- Capture device – an identifier for the capture device.
- Multi-modal data structures – in multi-modal systems, multiple data structures will exist for each modality.

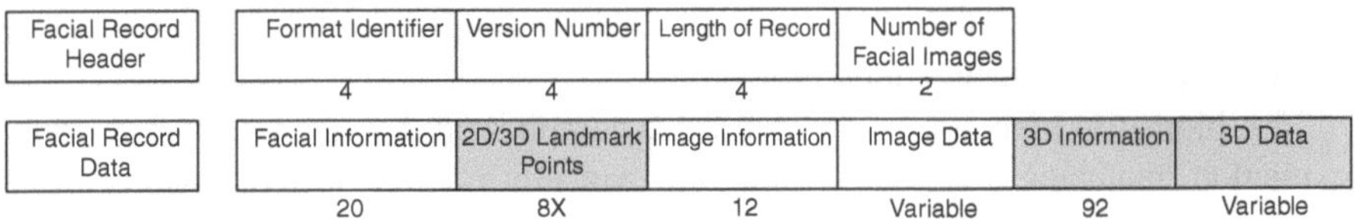

Fig. 7.2 Face image record format: overview (ISO19794-5 2009)

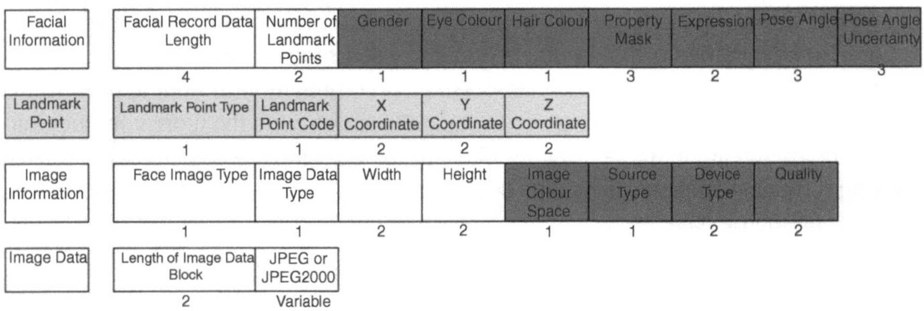

Fig. 7.3 Face image record format: facial record data (ISO19794-5 2009)

A key deliverable for these biometric standards is not to reduce the amount of or make public propriety information – for instance, how the extraction and matching algorithms of particular vendors operate. When stating what feature extraction and matching algorithms are used, it is necessary for the data structure to include information regarding these processes, in case subsequent components depend upon it to make a decision. For instance, in a system that uses multiple matching systems, the decision system needs to understand what results apply to which classification algorithms. It is important, however, that vendors have the opportunity to develop novel mechanisms without fear of losing competitive advantage, and standards are developed carefully to ensure this (versus levels of interoperability).

It is out of the scope of this text to describe the data interchange format for each modality (and the reader is guided to the individual standards as depicted in Table 7.2 for more information); however, to provide an understanding of the format an example will be given. Figure 7.2 illustrates the face image record format taken from ISO 19794-5 (2009). The record format is comprised of two components: Facial Record Header and Facial Record Data. There can be multiples of the latter but only one of the former. This allows for the communication of multiple images, through additional record data blocks. The primary purpose of the header is to provide basic information regarding how many images are present and the total length of the record.

Each Facial Record Data block is comprised of three mandatory (highlighted in white) and three optional blocks of data. Figure 7.3 provides a further breakdown of what each section actually stores (the numbers below the boxes indicating the size in bytes). Within this structure, a number of the values can be unspecified (indicated in dark grey). Information regarding number of landmark points, gender, eye colour and pose angle are included within the Facial Information block. The Image Information block contains information about the face image type (basic, frontal, full-frontal), dimensions of the image, sensor type and quality. The Image Data block contains the actual image data. The use of optional and unspecified entries allows for a greater flexibility in situations where such information is necessary.

Each part of the ISO 19794 standard contains a similar approach to that of facial recognition, but providing for the nuances required for each modality. Providing structure and context to the data format enables other components of the system to understand and interpret the data, removing the need for individual vendors to develop their own propriety formats and encouraging interoperability.

7.4 Data Structure Standards

Whilst the Data Interchange Format standards provide a mechanism for structuring the biometric data into a meaningful form, they do not provide any means of packaging this data for communication between systems or components. The Data Structure Standards do this by providing the necessary wrapper around the biometric data, referred to as the *Common Biometric Exchange Framework*. The ISO/IEC 19785 standard (in its four parts) defines the format and operation for the exchange of biometric data, as illustrated in Table 7.3.

The principal concept of the approach is to define a *Biometric Information Record (BIR)*. The structure of the record is comprised of three components, as illustrated in Fig. 7.4. The Standard Biometric Header (SBH) provides information to an application regarding the format and other properties of the *Biometric Data Block (BDB)*. The BDB contains the biometric data conforming to a defined format. The optional *Security Block (SB)* contains information pertaining to the encryption of BDBs within a BIR and the integrity of the BIR. A BIR may have one or more BDBs – so that multiple biometric samples can be packaged together. The purpose of the SBH is to provide a level of abstraction away from the BDB so that applications do not have to query the BDB to establish whether it is a sample that it wants to process. All significant information regarding the BDB(s) is contained with the

Table 7.3 ISO/IEC common biometric exchange formats framework

ISO/IEC reference	Description
ISO/IEC 19785-1:2006	Part 1: data element specification
ISO/IEC 19785-2:2006	Part 2: procedures for the operation of the biometric registration authority
ISO/IEC 19785-3:2007	Part 3: patron format specifications
ISO/IEC FCD 19785-4	Part 4: security block format specifications

Fig. 7.4 A simple BIR (ISO 19785-1 2006c)

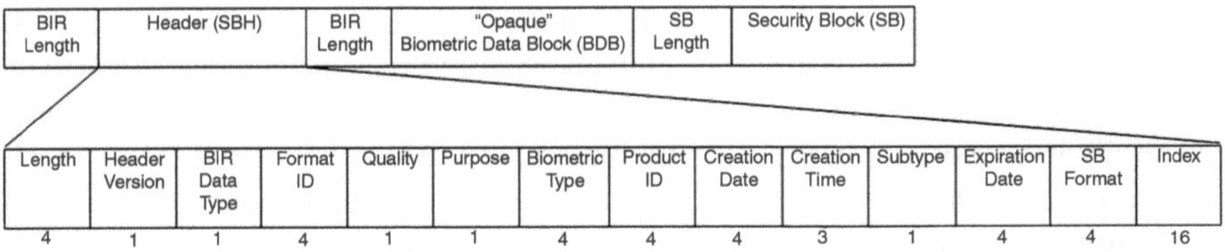

Fig. 7.5 BioAPI patron format

SBH. This approach also allows for additional security in terms of confidentiality and privacy of the BDB and integrity of the BIR. This particular standard allows for the BDB to be defined by vendors if required; however, it is anticipated that the previously presented Data Interchange Format standard would be utilised.

The standard also allows for the SBH to be defined to suit specific application scenarios that might arise. The standard refers to these as *CBEFF Patron Formats*. A CBEFF Patron is an organisation that intends to specify one or more CBEFF patron formats in an open and public manner. Patron formats for machine-readable travel documents, smartcards and XML all exist. The BioAPI Patron Format also exists, which is closely integrated with the Technical Interface Standards (refer to the next section). As depicted in Fig. 7.5, the BioAPI Patron format specifies the value and size of each data element with the SBH.

The length defines the size in bytes of the BIR, the BIR Data Type, whether encryption and integrity options are set and the processed level of the biometric sample – raw, intermediate, processed (or not available). The Quality and Purpose bytes indicate the quality of the BDB sample and its purpose, whether it is to be used for verification, identification, enrolment or audit. Biometric type refers to the modality and subtype to a specific form, such as the left or right hand, middle finger or thumb. As previously mentioned the BIR supports multiple BDBs and to facilitate this a complex CBEFF BIR structure exists. Effectively this utilises a nested form of the SBHs – please refer to ISO/IEC 19785-1 (2006c) for more information.

7.5 Technical Interface Standards

The Technical Interface standards provide an Application Programming Interface (API) so that various components of the biometric system are able to communicate, query and execute commands to each other. The Data Structure standards define the format for the BIR so that components can understand and interpret records and the Data Interchange Format standards provide the mechanism for extraction, matching and decision subsystems to extract the biometric information required. The Technical Interface standards utilise these to facilitate the movement, processing and storage of samples between components. The standards pertaining to this are presented in Table 7.4.

Part 1 of the standard specifies a high-level authentication framework that permits components from different vendors to interoperate through an API. The architecture,

Table 7.4 ISO/IEC Biometric programming interface (BioAPI)

ISO/IEC reference	Description
ISO/IEC 19784-1:2006	Part 1: BioAPI specification
ISO/IEC 19784-1:2006 Amd 1:2007	BioGUI specification
ISO/IEC 19784-1:2006 Amd 2:2009	Framework-free BioAPI
ISO/IEC 19784-1:2006 Amd 3:2010	Support for the interchange of certificates and security assertions and other security aspects
ISO/IEC 19784-2:2007	Part 2: biometric archive function provider interface
ISO/IEC FDIS 19784-4	Part 4: biometric sensor function provider interface

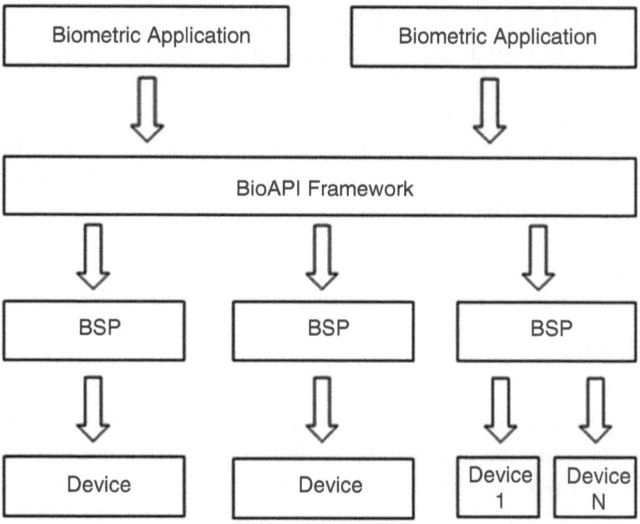

Fig. 7.6 BioAPI architecture

illustrated in Fig. 7.6, consists of a series of layers. The biometric application communicates to the BioAPI framework through a defined set of API calls (as described by this standard). The framework then invokes one or more Biometric Service Providers (BSPs) through a Service Provider Interface (SPI). Each BSP can be provided by different vendors and independently loaded as required by the application. The bottom layer of the architecture consists of the hardware and software components that perform the key biometric functions, such as capture or matching. Interactions between components from different vendors within the BioAPI framework can only take place if the data structures conform to international standards such as ISO 19785-1 (2006c).

The lowest level of the BioAPI consists of pieces of hardware and software utilised to perform biometric functions such as capture, archive, processing or matching. These components can be combined and controlled via a BSP or supplied as a separate BioAPI Function Provider (BFP). Through the use of common data standards, components are able to communicate using standardised data records. In this manner, a biometric application can compromise a variety of components supplied by different vendors, but that control and exchange information in a standardised manner – enabling system-level interoperability.

The BioAPI specification addresses the key functionality required within a biometric system. It consists of higher-level function calls, such as *BioAPI_Enrol, BioAPI_Verify* and *BioAPI_Identify*, and primitive functions, such as *BioAPI_Capture, BioAPI_Process, BioAPI_VerifyMatch* and *BioAPI_IdenitfyMatch.* The higher-level function calls encapsulate the necessary functionality to capture, create a template, process the sample, extract the features and perform the matching. The primitive function calls provide a finer granularity of control and enable each aspect of the process to be managed independently.

It is worth highlighting that BioAPI does not explicitly support multi-modal biometrics; however, as long as all appropriate components within the system are able to support complex BIRs this will function. Moreover, it is possible with some implementations to leave the management of the multiple samples to the biometric application, thereby effectively presenting a uni-modal approach to the BioAPI framework. This is useful when additional decisions, such as the current context of use versus the similarity score, need to be made.

7.6 Summary

Standards are key to ensuring interoperability within and between systems. Whilst some initial concerns might exist for vendors over developing standards, largely focused upon a misperception of loss of revenue, standardisation is in fact a key driver in the uptake of systems, providing customers with choice and providing an opportunity for all vendors to compete on an equal basis. More importantly, the development of standards also demonstrates a key milestone in the maturity of the domain.

The standardisation of biometric data samples, encapsulation of data and an API framework that permits fully interoperable biometric components is essential to providing more robust composite authentication. Whilst the BioAPI architecture was primarily developed for traditional point-of-entry applications, the functionality of the architecture and the interoperable nature of the components provide the enabling infrastructure to facilitate continuous and transparent identity verification.

References

ISO: ISO/IEC 19784-1:2006 Information technology – biometric application programming interface – part 1: bioAPI specification. International Standards Organisation. Available at: http://www.iso.org/iso/iso_catalogue/catalogue_tc/catalogue_detail.htm?csnumber=33922 (2006a). Accessed 10 Apr 2011

ISO: ISO/IEC 19794-1:2006 Information technology – biometric data interchange formats – part 1: framework. International Standards Organisation. Available at: http://www.iso.org/iso/iso_catalogue/catalogue_tc/catalogue_detail.htm?csnumber=38745 (2006b). Accessed 10 Apr 2011

ISO: ISO/IEC 19785-1:2006 Information technology – common biometric exchange formats framework – part 1: data element specification. International Standards Organisation. Available at: http://www.iso.org/iso/iso_catalogue/catalogue_tc/catalogue_detail.htm?csnumber=41047 (2006c). Accessed 10 Apr 2011

ISO: ISO/IEC 7501-1:2008 Identification cards – machine readable travel documents – part 1: machine readable passport. International Standards Organisation. Available at: http://www.iso.org/iso/iso_catalogue/catalogue_tc/catalogue_detail.htm?csnumber=45562 (2008). Accessed 10 Apr 2011

ISO: ISO/IEC 19794-5:2009 Information technology – biometric data interchange formats – part 5: face image data. International Standards Organisation. Available at: http://www.iso.org/iso/iso_catalogue/catalogue_tc/catalogue_detail.htm?csnumber=38749 (2009). Accessed 10 Apr 2011

Part III
System Design, Development and Implementation Considerations

Chapter 8
Theoretical Requirements of a Transparent Authentication System

8.1 Introduction

Part 1 of the book clearly identified the current issues surrounding user authentication, the increasing burden placed upon the end-user and the fundamental flaw that links authentication with an initial login rather than the specific action that carries the risk. It introduced the concept of transparent and continuous authentication as a process for enabling a more reliable and realistic measure of the system's confidence in the identity of the user. Part 2 of the book provided a more technical insight into the current authentication processes and illustrated how transparent authentication could function on an individual technique perspective. Key to this concept is the use of biometrics and this section also evidenced the role multibiometrics and biometric standards have in establishing effective transparent authentication.

The purpose of Part 3, the final part of the book, is to introduce the architectural concepts required to develop a Transparent Authentication System (TAS). It will provide a detailed commentary on what such a system would look like and the operational considerations that need to be taken into account. This chapter in particular will consider the theoretical requirements, architectural paradigms and provide an example of a TAS design.

8.2 Transparent Authentication System

The need to reconsider the nature of authentication has forced a redesign of a fundamentally flawed system of point-of-entry control. The authentication requirements, identified in Sect. 3.4, stated the need to do the following:

1. Reduce the authentication burden upon the user
2. Improve the level of security being provided
3. More closely link authentication of the user with the subsequent request for access

N. Clarke, *Transparent User Authentication: Biometrics, RFID and Behavioural Profiling*, DOI 10.1007/978-0-85729-805-8_8,

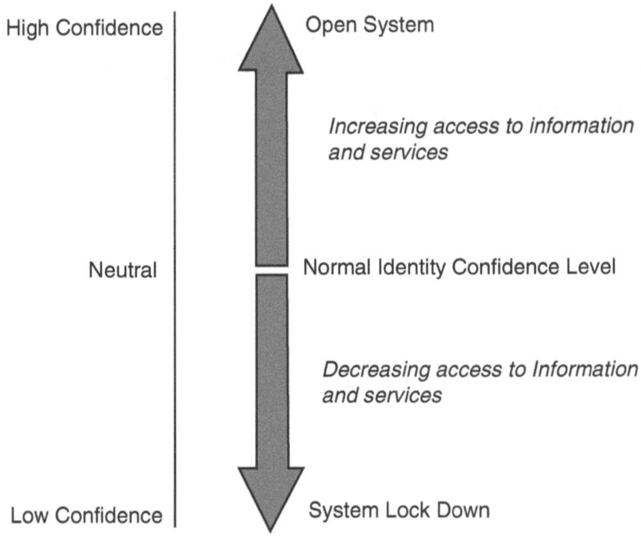

Fig. 8.1 Identity confidence

4. Take into account that a commensurate authentication approach is utilised depending upon the risk associated with the access request
5. Provide a more effective measure of identity confidence that goes beyond a simple Boolean decision

Whilst core to these requirements is the need to establish transparent authentication, the techniques in themselves, either on their own or together, do not enable a more robust and reliable means of user authentication, they are merely a component of a wider framework. In order to achieve these requirements, a framework that enables more intelligent handling of the authentication process is required. A TAS is a framework for providing a transparent, continuous, risk-aware, user-convenient and robust authentication system.

The fundamental reason for enabling transparent and continuous authentication is to provide the TAS with the ability to understand what level of confidence it has in the authenticity of the user on a continuous basis, not simply at point-of-entry but at all times. From that understanding it can determine whether to provide the user with direct and immediate access to services/applications requested, or require further verification. Figure 8.1 conceptually illustrates the process of the Identity Confidence – with a high confidence level the system is effectively open for the user. With a lower confidence, the access to services and applications are gradually restricted until the system is effectively locked down. The *Identity Confidence* will not only vary based upon the success or rejection of samples but will also degrade based upon a function of time. Therefore, if a user stops using the device, the Identity Confidence will not remain at its current level but will degrade with time until the device is locked down or the user provides additional samples. This will ensure potential misuse of the device is kept to a minimum.

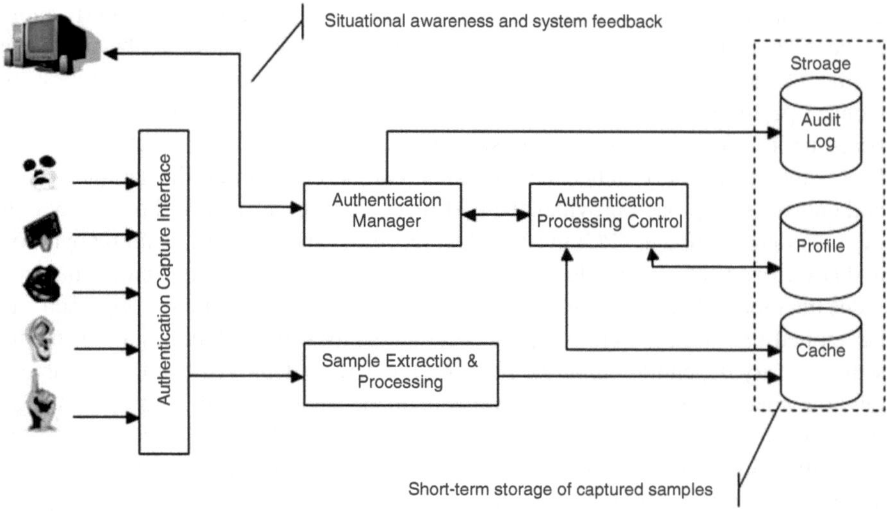

Fig. 8.2 A generic TAS framework

When the confidence is low, or a user selects an application requiring a higher confidence level, the TAS needs to establish the authenticity of the user. As transparent approaches have been ineffective to this point (this could be for a variety of reasons, such as a period of interactivity from the user and then suddenly requesting a secure service – rather than any failing of the TAS), an intrusive authentication request will be made to the user. Whilst this goes against the underpinning ethos, it is a necessary step required on occasions when the Identity Confidence is unable to provide a sufficient basis for the TAS to allow access. To facilitate this process, the suite of authentication techniques available to the TAS will also indicate in what mode they are able to operate: transparent, intrusive or both. Intrusive authentication, whilst placing a burden upon the user, will utilise biometric approaches to minimise the impact. It is envisaged that, in the normal operation of the system, sufficient transparent samples will be captured to maintain the Identity Confidence so that intrusive authentication occurs rarely or in situations where the particular service or application requires a very high level of authentication security.

From a high-level perspective, the operation of a TAS consists of five principal components:

1. Capture of authentication samples
2. Processing of authentication samples
3. Short- and long-term data repositories
4. Authentication Manager
5. Response

As illustrated in Fig. 8.2, the system enables the continuous transparent capture of authentication samples by the underlying system. Depending upon the particular configuration, this can enable capturing of samples during every user interaction or fewer if necessary. The samples are then stored in a short-term repository until the control interface, referred to as the Authentication Manager, decides to perform an authentication to verify or increase its confidence in the identity of the user. The Authentication Manager selects the most appropriate sample and the Processing Control performs the verification. The most appropriate sample can be determined based upon a variety of factors. Considerations, such as, selecting a sample that has been captured recently rather than an older sample or selecting the sample whose authentication approach provides the best recognition performance. For example, if on a personal computer (PC) the TAS had recently captured both facial recognition and keystroke analysis samples, it would be sensible to select the technique that provides the more reliable authentication result. Or indeed, a multi-modal approach can be applied that utilises both samples. What authentication techniques are available will also vary depending upon the platform the TAS is deployed upon – fundamentally on what hardware sensors are available on the device to enable transparent capture. From a TAS perspective, it will have access to a suite of authentication techniques that will include biometric, token and even secret-knowledge approaches. The latter approach is merely optional for situations where the hardware is unable to support many other authentication approaches. It also provides a level of flexibility within the framework to allow for such techniques – particularly if a system designer or end-user has a particular preference towards a form of secret-knowledge approach.

Based upon the authentication result, the Authentication Manager will modify its confidence. At some point, it may be necessary to invoke an intrusive authentication request, and the Response mechanism is required to provide the necessary interface and associated information to the user. The long-term repository is utilised to store the biometric templates and other authentication data. This could include previous biometric samples to prevent replay attacks or just the samples that were successfully authenticated so that they can be used within a template renewal process.

The final key input into the TAS is the system status. In order to provide commensurate authentication security, the TAS needs to obtain a situational awareness of what the user is doing. Therefore it is necessary for TAS to integrate into the OS in such a manner as to act as a checkpoint before the reference monitor. Figure 8.3 illustrates how a TAS provides a confirmation of authenticity prior to the access control decision being made by the reference monitor. It is not the purpose of the TAS to provide access control or maintain the list of access control privileges. This process is handled by the access control mechanisms that reside within the operating system (OS) of the host device. In this manner, the TAS is able to ensure that an appropriate level of Identity Confidence exists prior to allowing access via the reference monitor.

The TAS framework provides the ability to understand what the continuous Identity Confidence is and provides the processes that interface with the suite of authentication techniques. It is, however, the authentication techniques that provide the basis for a successful TAS. The strength of protection provided by the TAS is wholly dependent upon the authentication techniques performing well. If the authentication suite comprises of techniques with poor levels of recognition performance, perhaps due to ineffective vendor algorithms, the resulting performance will also be poor. As highlighted in the previous chapter, transparent authentication can be achieved in a variety of ways – each individual case depending upon the availability

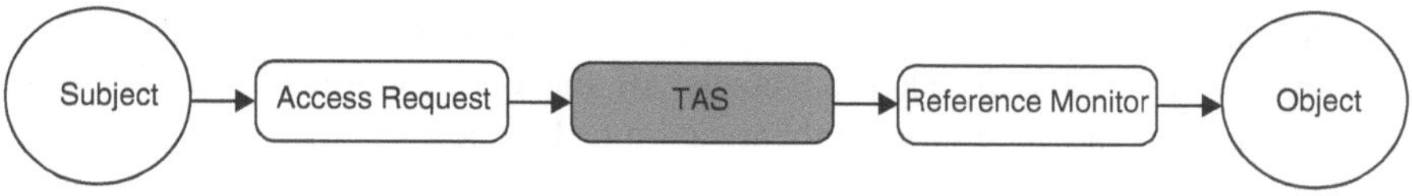

Fig. 8.3 TAS integration with system security

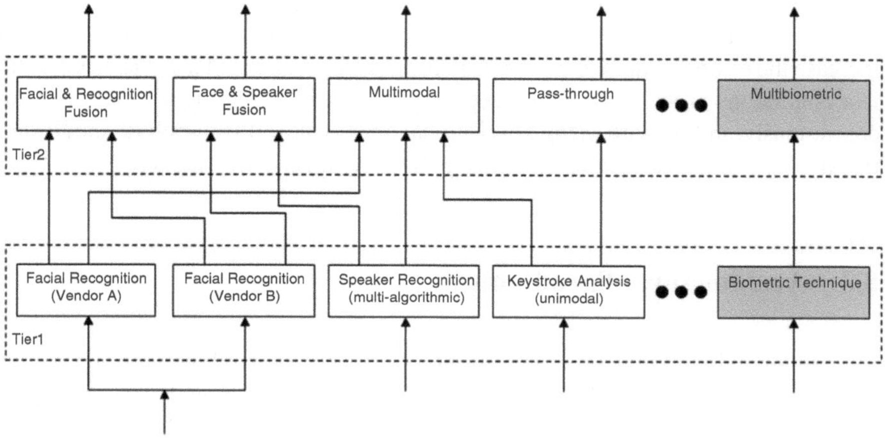

Fig. 8.4 Two-tier authentication approach

of sensor hardware on the technology platform. However, increased protection comes via the use of multibiometric-based approaches. In order to provide a degree of flexibility, a TAS takes a two-tier approach to the process, as illustrated in Fig. 8.4. Tier 1 is the core authentication technique provided by the biometric vendor. This could simply be a uni-modal technique or incorporate any arrangement or configuration of multi-modal, multi-algorithmic, multi-instance, multi-sample approaches. When possible, the output of tier 1 techniques is the match-level score (rather than the decision-level result). The optional tier 2 provides an opportunity for an independent multibiometrics approach. These tier 2 modules capitalise upon tier 1 outputs and are able to utilise a complete range of multibiometric approaches. The advantages of using this second tier are the following:

- A lack of vendor-specific multi-modal approaches exist that combine the necessary biometric techniques that are available within the TAS. The variability of authentication techniques available with each hardware device tends to exasperate the problem.
- The possible permutations and combinations of techniques provides for a large number of variations, some multi-modal, others multi-algorithmic, etc. Rather than multiplying the number of classifiers required to perform the underlying technique matching, the two-tier approach allows for a separation and simplification of the process.
- Tier 2 provides the ability to specifically design fusion-based classifiers that are independent of the base authentication techniques.

The use of a two-tier structure provides for the highest degree of flexibility within the framework and thus takes advantage of the optimisation and performance enhancement of the underlying authentication techniques.

The TAS approach also provides an almost natural anti-spoofing safeguard. As an attacker would not know in advance which samples are being used in the authentication, they have no choice but to attempt to spoof each and every approach a device can support. Whilst each individual technique is no more difficult to spoof than normal, the need to spoof several techniques sometimes simultaneously will increase the effort required to circumvent the control. Furthermore, as the mechanism is a continuous process, it is not merely a process of fooling the sensors once but continuously. In line with many multi-modal approaches, it is therefore anticipated that a TAS will provide far more robust protection against directed attack.

8.3 Architectural Paradigms

The implementation of a TAS does introduce a variety of additional considerations that play an important role over the viability of such an approach. The TAS itself is independent of any architectural constraints, but the realisation of the system will require a decision to be made over the architectural structure. Fundamentally, three approaches to implementation exist: network-based, host-based or hybrid – a combination of both network and host paradigms. From an architectural perspective, both principal approaches have their own set of advantages and disadvantages. And these considerations include the following:

- The inclusion of multiple authentication approaches, particularly biometric-based, can introduce a significant impact upon systems resources, especially for very computational-constrained devices.
- The need to utilise a range of vendor-supplied biometric algorithms might make it more cost effective to be performed centrally rather than individual licenses.
- The development, management and maintenance of various biometric approaches across a range of technologies will also impact upon what biometric vendors will be able to support. Central management provides platform independence and enables vendor algorithms to be updated in a timely fashion.
- Large populations will quickly impact the network and computer resources required for central management.
- Privacy concerns over the storage of biometric template information and who has responsibility could be issues for consideration; end-users might have concerns over central storage and the ability of an organisation to handle their data in a trustworthy fashion. Storing the templates locally also provides attackers with direct access to the template files (assuming sufficient access).
- Timeliness of system response: central management will introduce a level of delay within TAS, as samples are sent centrally to be processed.

A network-centric or centrally managed approach will direct all the key computational tasks and storage to a central server. For TAS implementations based upon an

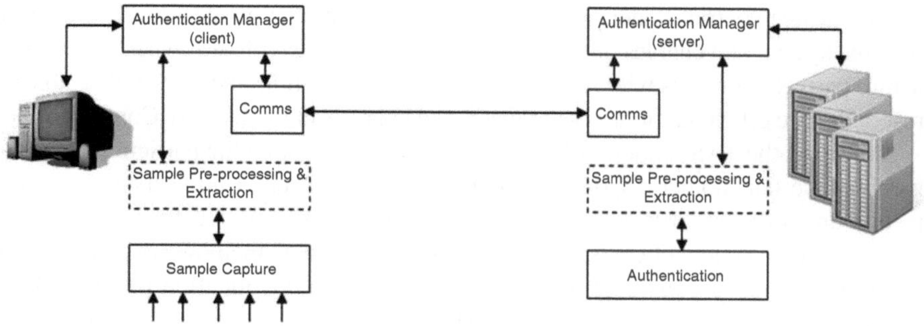

Fig. 8.5 Network-centric TAS model

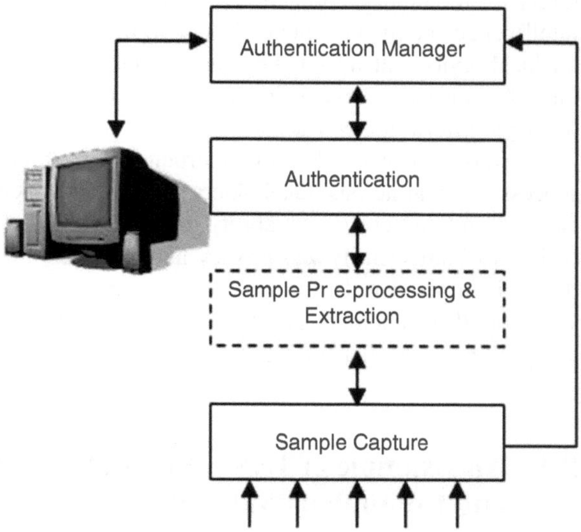

Fig. 8.6 A device-centric TAS model

organisation, the central management and facilitation of the system provides for a far greater granularity of control, with management interfaces able to demonstrate real-time system information. A third-party–managed authentication services company would also benefit from a network-centric model. Other stakeholders such as network operators may also have a valid argument for wanting to retain the intelligence within the network. It disassociates much of the overhead and valuable information from end points and provides greater oversight from an accountability perspective. As illustrated in Fig. 8.5, the mobile device itself will act as the authentication/biometric capturing device and be able to respond to a decision sent from the server to permit or restrict access to a user. But the processing and verification of the authentication samples, alongside the intelligence in the TAS, are hosted centrally.

In a device-centric approach the whole TAS is deployed on the device. All the information, algorithms and management controls required for the authentication processes are stored on the device. Furthermore, all the processing required to perform the verification also takes place on the device. Figure 8.6 illustrates an example of such an approach.

The device-centric model provides the end-user with the greatest level of control over the system and enables biometric-based data to reside with the owner, rather than a third party. For devices with sufficient computational abilities the TAS will be able to respond more efficiently to access requests and changes in the Identity Confidence.

In reality, the TAS model utilised will largely depend upon the particular requirements of the stakeholder responsible for deploying the system. End-users not wanting to pay any support costs to a third-party may pay a one-off fee for a standalone device-centric version. Organisations may wish to implement their own network-based solution, or indeed outsource the task to a specialist authentication provider. It is envisaged in the majority of situations that the hybrid model presents a compromise between the two approaches that satisfy more of the general requirements. A hybrid model will enable some authentication processing locally but also provide an opportunity for the more computationally extensive tasks to be completed centrally. The intelligence and management of the TAS in a hybrid model could reside with either the local device with a feed of management information back to a central management interface, or central management, with related information being sent back to the device to inform the end-user.

The topology of a TAS is an important factor to consider at the outset of the design process. With numerous stakeholders (such as network operators, corporate IT administrators and end-users), the ability to provide identity verification in a manner that maintains both security and privacy and considers the operational impact upon the device is imperative. Unfortunately, however, it is difficult to maintain all these services for all stakeholders, and a trade-off frequently exists between different security and privacy issues, depending upon what the system is trying to optimally achieve.

8.4 An Example of TAS – NICA (Non-Intrusive and Continuous Authentication)

In order to provide a more detailed appreciation of a TAS, the following example is presented based upon research conducted by the author, funded by a major UK mobile network operator (Orange PCS) and further enhanced by a grant from the EduServ Foundation (Clarke 2004; Clarke and Furnell 2006; Karatzouni et al. 2007). This will also enable a further insight and discussion on various additional aspects of a TAS that result when considering its design.

The focus of the research was to design and implement a TAS for use within mobile phones. From earlier research, it was identified that whilst the network operator had preferred a centrally managed solution, end-users opted for a device-centric model. As such, the TAS developed was based upon a hybrid model. The resulting model was referred to as *NICA – Non-Intrusive and Continuous Authentication*. The architecture for the system is illustrated in Figs. 8.7 and 8.8. The architecture consists of a number of key TAS components, such as an ability to capture and authenticate biometric samples, an intelligent controller, administrative capabilities and the storage of biometric profiles and authentication algorithms. The system was designed with

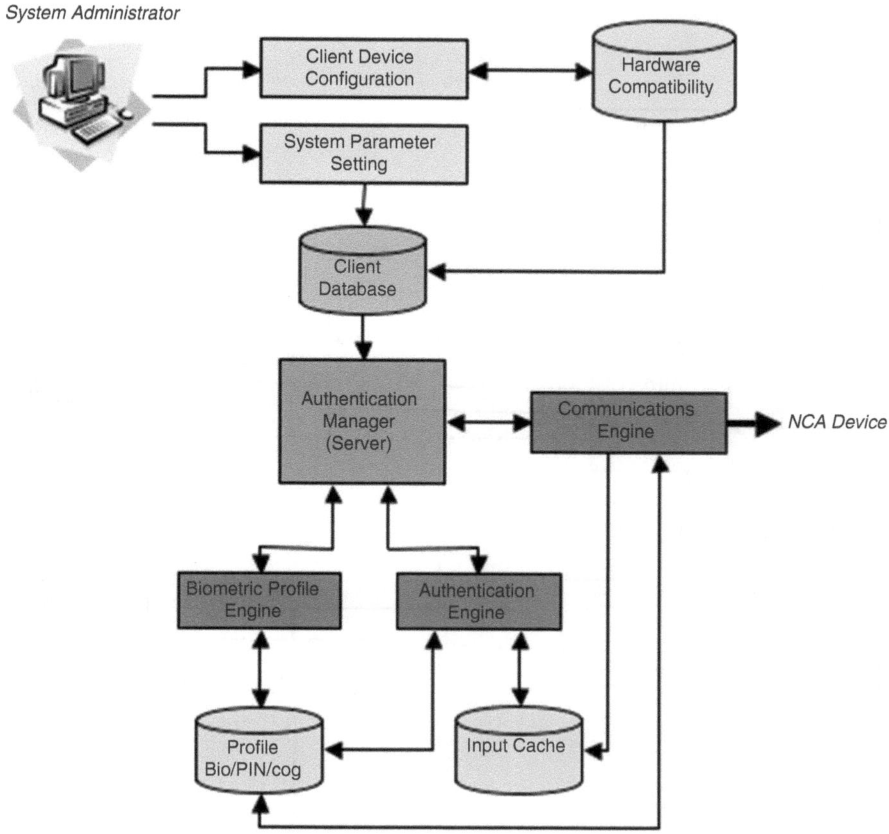

Fig. 8.7 NICA – server architecture

the flexibility of operating in an autonomous mode to ensure security is maintained even during periods with limited or no network connectivity.

The majority of the device topology, as illustrated in Fig. 8.8, is identical to the server architecture, with the operation of the process engines, storage elements and Authentication Manager remaining (in principle) the same. The device topology does, however, introduce a number of additional components that provide the input and output functions of the system. The fourth process engine in the form of the Data Collection Engine is included on the device topology and provides the input mechanism, which collects and processes users' device interactions. The output components consist of an Intrusion Interface and Security Status. The former provides the NICA to OS connection for restricting user access and provides user information as and when required, and the latter provides an overview to the Identity Confidence and security of the device.

The implementation of the architecture will differ depending upon the context that a device is being used within. For instance, in a standalone implementation the device has no use for the Communications Engine – as no network exists to which it can connect. Meanwhile, in a client–server topology the components required will

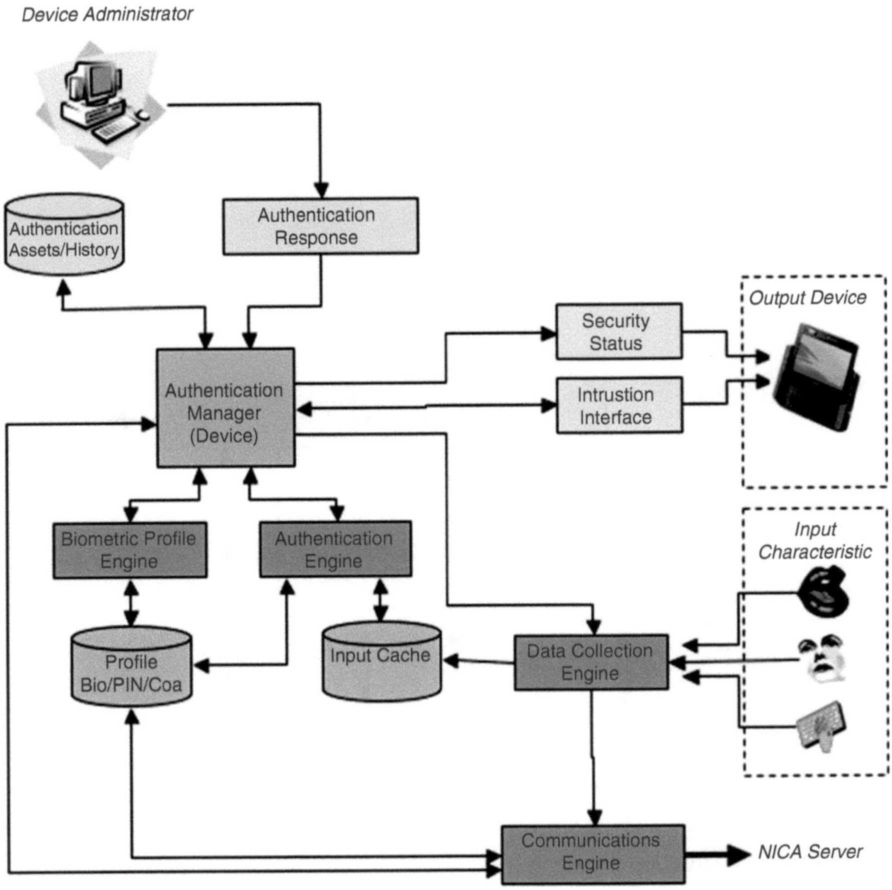

Fig. 8.8 NICA – client architecture

vary depending upon the processing split between the server and client. There are numerous reasons why a network administrator may wish to split the processing and control of NICA differently, such as network bandwidth and availability, centralised storage and processing of the biometric templates and memory requirements of the mobile device. For example, in order to minimise network traffic, the network administrator may require the host device to authenticate user samples locally or, conversely, the administrator may wish the device to only perform pre-processing of input samples and allow the server to perform the authentication, thus removing the majority of the computational overhead from the device, but still reducing the sample size before transmitting across the network.

The framework operates by initially providing a baseline level of security, using secret knowledge approaches, which progressively increases as the user interacts with their device and biometric samples are captured. Although user authentication will begin rather intrusively (e.g. when the device is switched on for the first time), with the user having to re-authenticate periodically, the system will quickly adapt, and as it does so, the reliance upon secret knowledge techniques is replaced by a reliance upon biometrics – where the user will be continuously and non-intrusively authenticated. The result is a highly modular framework that can utilise a wide range of standardised biometrics, and which is able to take advantage of the different hardware configurations of mobile devices – where a combination of cameras, microphones, keypads, etc. can be found.

8.4.1 Process Engines

The computational backbone of this framework, from template generation to file synchronisation is provided by four process engines:

• Data Collection Engine	• Authentication Engine
• Biometric Profile Engine	• Communications Engine

The Data Collection Engine is required to provide the capturing of a user's input interactions. The actual samples to be captured by the engine will be dependent upon the hardware contained within the device – as the hardware configuration of devices can vary considerably, with a mixture of cameras, microphones, keypads and even fingerprint sensors on some devices. However, as an example, the typical hardware configurations on a typical mobile handset will permit the framework to capture the following types of input sample:

- Keystroke latencies from a keypad whilst the user is typing – including entering telephone numbers and the typing of text messages
- Images from a camera whilst the user is interacting with the device
- Sound from a microphone whilst the user is voice dialling

The Data Collection Engine, as illustrated in Fig. 8.9, obtains the input data through a number of OS-dependent interfaces, one for each type of input data, and a system monitor. With the exception of the system interface, each of the interfaces captures and logs samples from their respective input devices. The system interface monitors device activity in order to establish which services and applications the user is accessing, so that the Authentication Manager can ensure the user has the correct authentication level, or Identity Confidence, required to access them.

Once the software interfaces have captured a user's interactions, the next stage of the Data Collection Engine is to pre-process this data into a biometric sample – thereby removing any erroneous data from the sample, leaving only the information that is required for the authentication. This stage is optional in the client–server topology, but recommended to minimise the size of the template file that has to be sent across the network – thereby reducing network traffic. The task of pre-processing (and other biometric-specific operations such as template generation and authentication) is dependent upon the individual biometric technique – each one will

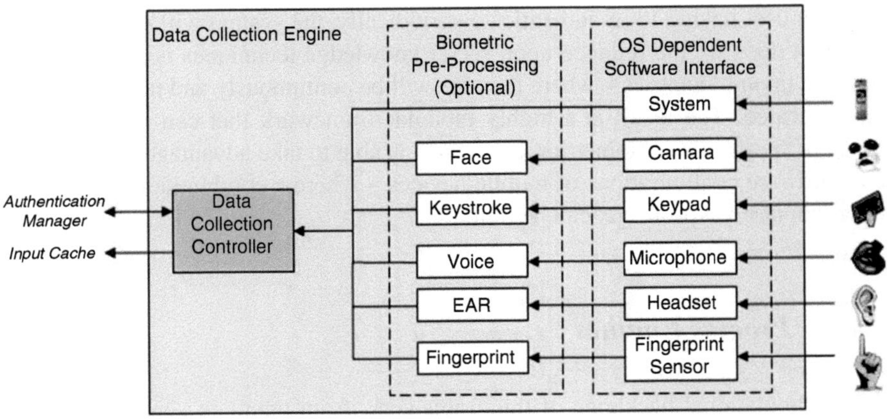

Fig. 8.9 NICA – data collection engine

pre-process the input data in a different way. For instance, pre-processing of keypad input data to be used by keystroke analysis involves the calculation and scaling of timing vectors, whereas the pre-processing of a sound file for use by a Voice Verification technique would involve the extraction of key sound characteristics.

In order to achieve modularity – so that the framework can utilise one or more biometric techniques in any given device – the system must be compatible with a wide range of different biometric techniques from a number of biometric vendors. Although all the algorithms associated with a biometric technique are proprietary, the biometric standards will be utilised to format, package and communicate all information necessary, requesting for sample pre-processing, template generation and authentication – thereby negating the requirement for any custom software development of biometric algorithms.

After the pre-processing, the Data Collection Controller will proceed to send a control message to the Authentication Manager, informing it what input data have been captured. This enables the manager to utilise the most appropriate input sample in a given situation. It will also send the biometric sample to the Input Cache for temporary storage and possible subsequent use in authenticating the user.

The Biometric Profile Engine's primary role is to generate the biometric templates that enable subsequent classification by the Authentication Engine. This is achieved through a series of template generation algorithms, which take the biometric sample and output a unique biometric template. The Biometric Profile Controller, as illustrated in Fig. 8.10, takes both the biometric sample and biometric template and stores them within the Profile Storage element – the template is utilised in the authentication mechanism and the sample can be used again in re-generating or retraining the biometric template. The data control interface to the Authentication Manager is bi-directional, as the Authentication Manager initiates profiling and the Controller communicates with the Authentication Assets table, via the Authentication Manager, updating the usability status of the respective techniques, thus enabling the Authentication Manager to subsequently utilise the technique should it wish to.

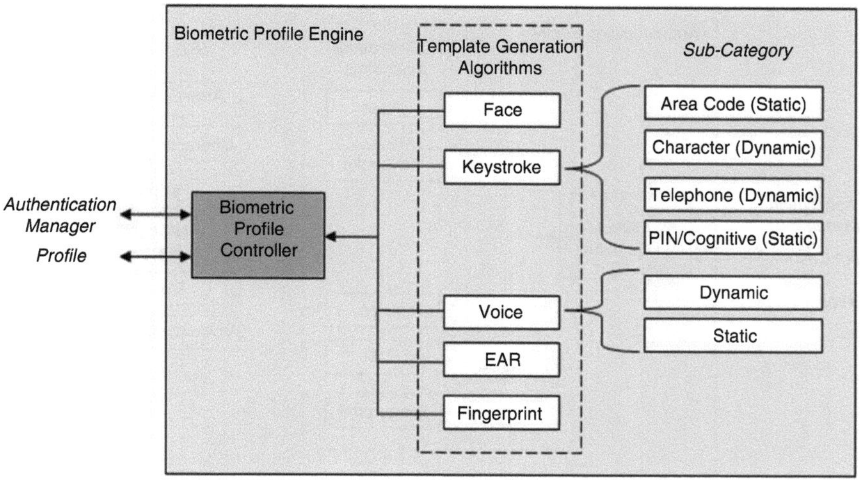

Fig. 8.10 NICA – biometric profile engine

As the framework utilises a number of biometric techniques (each of which can themselves have many different input types) it raises the problem of template generation. It would be implausible for the system to require the user to provide all the input data required to generate the templates at initial device registration – especially keystroke analysis, where the template generation algorithm typically requires a number of samples before the template can be created. This problem is solved due to the philosophy behind the framework itself – which initially provides no security beyond what is currently available on mobile devices (i.e. secret-knowledge approaches such as the personal identification number (PIN) or password) and for the biometric security to gradually increase as the user naturally interacts with their device. Apart from initially setting the PIN code and data for other secret-knowledge approaches (e.g. cognitive passwords), this will remove any requirement to generate any templates (which, given the number of input scenarios, could take a significant period of time to setup), with the secondary and subsequent effect of ensuring the input data utilised in the creation of templates are truly representative of the natural user's device interaction. However, it is likely that some biometric templates may be generated during device registration, such as facial recognition. If the device has the hardware available, several images of the user can be taken whilst the user is setting the cognitive responses and can be subsequently used to generate the template.

The Authentication Engine is controlled by the Authentication Manager, and performs the actual authentication. When requested, the Authentication Engine will perform the authentication by retrieving the required input data from the Input Cache and the corresponding biometric template from the Profile Cache, as illustrated in Fig. 8.11. If the Authentication Engine approves the sample, the data are moved from the Input Cache to the Profile, for use in subsequent retraining. If the sample fails, it is deleted from the Input Cache.

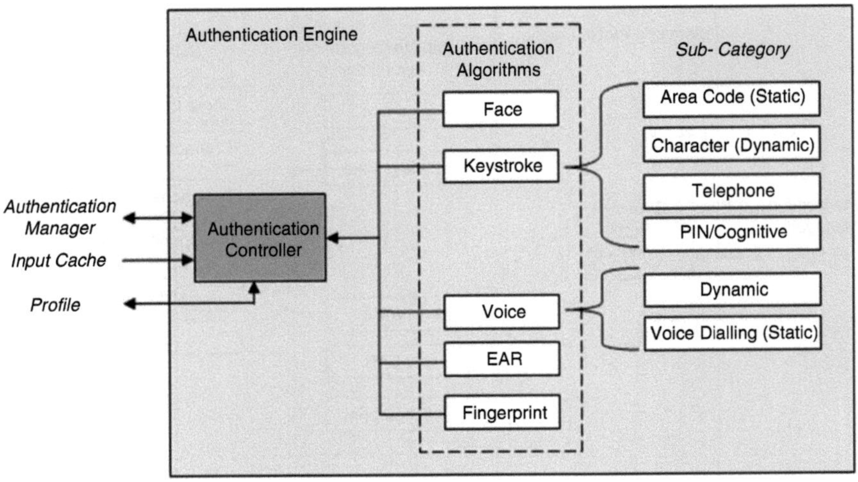

Fig. 8.11 NICA – authentication engine

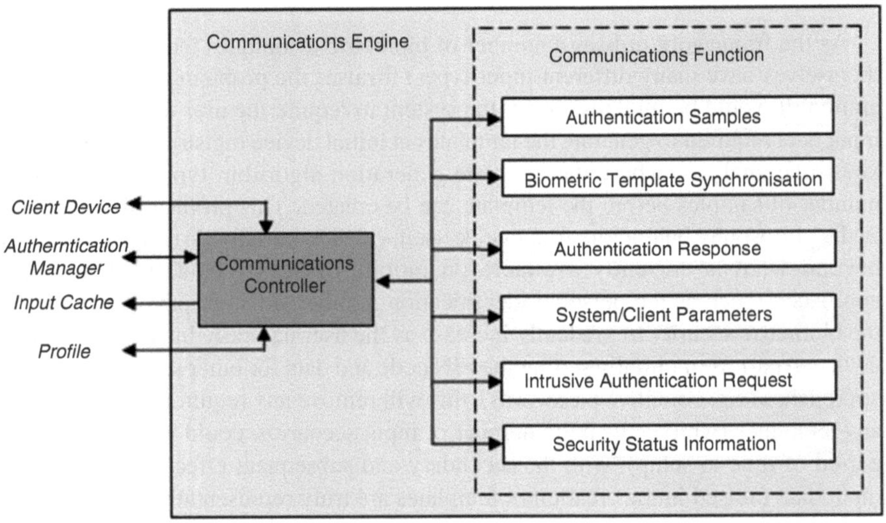

Fig. 8.12 NICA – communication engine

The result of the authentication is reported to the Authentication Manager, with the subsequent response being dependent upon a decision process based upon variables such as the Identity Confidence, recognition performance of the techniques and the status of the system.

The Communications Engine provides a means of connecting server and client. The actual mechanism by which this is achieved is independent upon the wireless access technology available to the mobile device, but is a function of the device rather than the authentication architecture. The role of the engine is to transfer information based upon six categories, as illustrated in Fig. 8.12.

The authentication samples will vary in nature depending upon the biometric techniques available to a particular mobile device. If the client–server topology is defined, where some or all of the biometric analysis is performed locally, then it is imperative that the Communications Engine performs biometric template synchronisation to ensure the most up-to-date and relevant templates are being employed. Authentication Response and System/Client parameters allow server and client to know what mode of operation they are working in, which side performs authentication and for which authentication techniques. For instance, it would be plausible to perform the less computational complex functions, such as password/cognitive response verification, on the mobile device, reducing network overhead and improving response time, but leave the server to perform the more complex biometric verifications. Although not defined by this mechanism, it is assumed the communication and storage of the biometric samples are performed securely to ensure no manipulation of the templates can occur. The Communications Engine also enables the server to provide feedback to the user, either through requesting they intrusively authenticate themselves or via a Security Status – a mechanism that enables the user to monitor the protection provided to the device.

8.4.2 System Components

This section addresses the various system components of the framework, which help to provide the system parameters and define the authentication techniques. This section is split into the following:

- Security Status and Intrusion Interface
- Authentication Assets
- System Administration and Response

The Security Status and Intrusion Interface components represent the two output processes of the framework. The Security Status simply provides information to the end-user regarding the level of security currently provided by the device, the success or failure of previous authentication requests and the Identity Confidence. Although it is perceived that many users will have no interest in viewing this, it is included as an informational guide to the user.

Each mobile device must be able to establish the level of security it can provide. This is achieved by determining which authentication techniques are currently enabled (i.e. have a template generated and/or password enabled). As the performance of biometric techniques varies, a process is required to ensure weaker authentication techniques do not compromise the level of security. To this end, each of the authentication techniques is given a confidence level depending upon its recognition performance. As illustrated in Table 8.1, the confidence levels are separated into two types, those concerned with biometric techniques and those based on secret-knowledge techniques. The biometric techniques are categorised on their published equal error rate (EER) for the system, with the system having more confidence the lower the false acceptance rate (FAR). The secret-knowledge techniques are split

Table 8.1 Confidence level definitions

Biometric		Secret knowledge	
Confidence level	EER level	Confidence level	Input required
B0	10–20%	S0	PIN/Cognitive
B1	5–10%	S1	PUK (operator)/Administrator password
B2	2–5%		
B3	0–2%		

Table 8.2 NICA – Authentication assets

ID	Technique	Technique sub category	Topology	Device/ network compatibility	Template gen date	Confidence level	Intrusive
1	Cognitive	Phrase #	Both	True	–	B2	True
2	EAR		–	–	–	B0	True
3	Face		–	–	–	B1	True
4	Finger		–	–	–	B0	True
5	Keystroke	Static #	Both	True	–	B2	False
6		Dynamic	Both	True	18/07/2007	B3	False
7	PIN		Both	True	12/07/2007	P0	True
8	Signature		–	–	–	B2	False
9	Voice	Static #	Server	True	15/07/2007	B2	False
10		Dynamic	Server	False	–	B3	True
:	:	:	:		:	:	:

into two levels: S0, which represents PIN, password or cognitive challenges, and S1, which represents the administrator password or the personal unlock key (PUK) code in cellular network terms.

The Intrusion Interface is required on the occasions when the identity of the user needs to be verified before access or continuation of services. Although the framework is designed to operate in a non-intrusive manner, there are occasions when the user will be required to authenticate himself or herself, perhaps when the system has already non-intrusively and unsuccessfully attempted to authenticate the user several times.

The Authentication Assets are a detailed breakdown of the authentication mechanisms available to the Authentication Manager for a particular mobile device, as illustrated in Table 8.2. The information contained within the Authentication Assets is used by the Authentication Manager to determine which techniques are supported and, in particular, which techniques have templates generated and what the corresponding confidence levels are for the technique. This enables the Manager to decide upon which technique, given the contents of the Input Cache, would be most appropriate for use in subsequent authentication. Within the client–server topology the Authentication Assets for a client are stored within the Client Database.

Given a wide range of mobile devices exist, each differing in terms of their hardware configuration and operating system, it becomes implausible to design an authentication mechanism that will automatically work on all devices. The difficulty is in enabling the framework to plug-in to the different operating systems, particularly the software interfaces of the Data Collection Engine and the Intrusion Interface. Although the framework is system- and device- independent in its approach to authentication security, the practical nature of utilising different operating systems would require a level of OS-dependent software design.

In a client–server topology, a network is very likely to consist of a varied number of different mobile devices to which the framework will have to adapt, ensuring a high level of security given any combination of hardware. To this end a Hardware Compatibility Database (HCD) is included in the server architecture to provide a list of all compatible devices, along with specific information about which authentication techniques are available for which device.

In order to achieve a commensurate level of security versus user convenience, for a given mobile device, the framework allows a device administrator to define a number of system parameters. These include:

- Enabling or disabling NICA (per device or per user)
- Individual enabling/disabling of authentication techniques
- Determining the processing split between client and server (in the network version only)
- Input Cache raw data period
- Profile storage of raw data period
- System Integrity period
- Manual template generation/retraining
- Monitoring Authentication Requests
- Monitoring System Integrity levels
- Security level – high, medium, low (has an effect on the threshold values chosen for the biometric techniques)
- Topology – standalone or client–server

In addition to the above parameters, the device administrator is also permitted to define the Authentication Response table. As shown in Table 8.3, this defines a number of key services that the device can access, including bank accounts, share dealing, micro-payments and expensive video calls with a corresponding Identity Confidence, Authentication Confidence Level and (where applicable) Location Access. In order for a user to access any of the services listed they must have a System Integrity level greater or equal to that specified. If not, the user will be subsequently required to authenticate himself or herself using an authentication technique with the corresponding Confidence Level in order to proceed with the service. If they are unable to obtain the required Confidence Level, the service will remain inaccessible to them.

The purpose of the Authentication Response table is to ensure key services and file locations are protected with a higher level of security than less private information or less risky services. Determining which service or what information the user is accessing at any particular moment is achieved via the Authentication Manager, in conjunction with the System Monitor within the Data Collection Engine.

For mobile devices without a B3 level of security, the administrator is left with two possibilities:

- To lower the Identity Confidence levels within the Authentication Response table appropriately, although this is not recommended, as it would lower the security of the device.

Table 8.3 NICA – Authentication response definitions

Service	System integrity level	Confidence level	Location access (if applicable)
Bank account access	=>4	B3	http://hsbc.co.uk
Downloading media content	=>3	B2	
Media message	=>1	B1	
Micropayments	=>4	B3	
Share dealing	=>4	B3	http://sharedeal.com
Text message	=>−1	B0	
Video call (international)	=>2	B2	
Video call (national)	=>1	B1	
Video call (other cellular networks)	=>2	B2	
Voice call (international)	=>2	B2	
Voice call (national)	=>0	B0	
Voice call (other cellular networks)	=>1	B1	
:	:	:	:

- To intrusively request the user to authenticate himself or herself each time he or she wishes to use the most sensitive services, using a cognitive response and keystroke analysis approach (otherwise known as a hardened password).

8.4.3 Authentication Manager

The Authentication Manager is the central controlling element of the framework, with control over the process engines and intrusion interface. The role of the Authentication Manager includes:

- Determining the topology and administrative setting
- Generating and maintaining the System Integrity level
- Requesting profile generation and retraining
- Making authentication requests, both intrusive and non-intrusive
- Determining what subsequent action should be taken, given the authentication result
- Determining whether a user has the required Identity Confidence level
- Requesting removal and lock down of services and file locations
- Use and maintenance of the Authentication Assets

However, the principal task of the Authentication Manager is to maintain the level of security required commensurate with the services being provided by the device. To this end, two key processes operate to ensure this:

- System Integrity
- Alert Level

The System Integrity is the realisation of the TAS Identity Confidence measure. The process assists in ensuring security through maintaining a continuous level of

Table 8.4 NICA – System integrity settings

Confidence level	Increment/ Decrement value	Maximum system integrity level
S1	None – System integrity set to 0	NA
S0	NA	NA
B3	2	5
B2	1.5	4
B1	1	3
B0	0.5	2

probability in the identity of the user. It is a sliding numerical value between −5 and +5, with –5 indicating a low security, 0 a normal 'device switch-on' level and +5 indicating a high security level. The System Integrity varies depending upon the results of authentication requests and the time that has elapsed between them. Each of the confidence levels are given a number which is added or subtracted from the System Integrity dependent upon whether the technique has passed or failed the input sample, up to a defined maximum level. This System Integrity level is a continuous measure, increasing and decreasing over the duration of a user's session. Table 8.4 illustrates by how much the System Integrity level is to be increased or decreased for each of the confidence levels. The maximum System Integrity level is included to ensure a user is unable to achieve the higher integrity levels by utilising techniques with relatively high false acceptance rates (i.e. those with lower confidence levels – B1, B0). This ensures a user with a System Integrity level of 5 has not only had consistent successful authentication requests during his or her session, but has also recently been authenticated by a biometric technique with a confidence value of B3.

The secret-knowledge confidence levels have a somewhat different role. S1 has the effect of resetting the System Integrity level – since in a client–server topology, S1 represents the network administrator overriding the password (or PUK code in cellular terms). In a standalone topology, typing S1 would either unlock the device or give access to the host administrative settings. The S0 level is only required in two scenarios – first, if no biometric approaches are available for use (in which case the Authentication Manager will resort to providing a basic level of security through the use of the PIN, password or cognitive response), and second, as a means of providing two-factor authentication – the PIN or password is used in conjunction with keystroke analysis to verify the authenticity of the user. In this special situation, whatever the confidence value of the respective keystroke analysis technique, if successful, the level is increased by 1 (up to the maximum of B3) – as this represents a multi-modal approach, referred to specifically as a hardened password (Monrose et al. 1999).

The period of time that has elapsed between authentication requests also affects the System Integrity level. In order to ensure that devices remaining unused for a period do not continue to have a high integrity level, which could be subsequently misused by an impostor to access more sensitive information and expensive services, the integrity level begins to decrease over time. The actual period is to be configurable

Table 8.5 NICA – Authentication manager security levels

Authentication security level	Description of alert
1	Normal
2	Authenticate on users next input
3	Request entry of a B3/B2 level or PIN/Cognitive question
4	Lock handset – requires unlock code from network operator

on a per user basis and ideally a function of usage. After each defined period of misuse the System Integrity level decreases until the normal security level of 0 is reached. In practical terms this means a mobile device with a period set to 30 min with the highest integrity level of 5 will take 2 h 30 min to decrease down to a normal integrity level. Negative System Integrity values, however, remain until a subsequent authentication request changes it (for the better or worse) or the S1 level code is entered.

The Alert Level is the second of the key security processes working at the core of this framework. The algorithm effectively implements a cascade multibiometric (although on occasions multi-authentication) approach, where the result of one authentication determines whether subsequent authentications are required. It, however, also includes a process of requesting intrusive authentication upon failure of the transparent cascade authentication. The Process algorithm has four stages of authentication security (depicted in Table 8.5 and illustrated in Fig. 8.13) with the level of authentication security being increased until the device is locked (requiring an administrative password or PUK code from a cellular network provider). The number of stages was determined by a compromise between requiring a good level of user convenience and better security. Through mixing transparent and intrusive authentication requests into a single algorithm it is expected that the majority of authorised users will only experience the transparent stages of the Process algorithm. The intrusive stages of the algorithm are required to ensure the validity of the user utilising the stronger authentication tools before finally locking the device from use. The difference between intrusive and transparent authentication requests is identified in the algorithm. The intrusive authentication requests require the user to provide a biometric sample of confidence value B3 or B2 (whichever is available and higher).

The general operation of the Authentication Manager is to periodically send an authentication request to the Authentication Engine; here, periodically is definable administratively, but would typically be a function of usage. The Authentication Engine will subsequently retrieve the last and highest (in terms of confidence value) set of user's inputs (i.e. a camera image from a video conference call or a sound file from voice dialling) from the last *x* minutes (where *x* is also definable administratively, but would typically be in the region of 2 min). If the Authentication Engine passes the input, the Authentication Manager goes back into a monitoring mode. If not, the Authentication Manager performs an authentication request again, but using the remaining data from the Input Cache from the last *y* minutes (where *y* is definable administratively, but would typically be in the region of 5 min) – using the

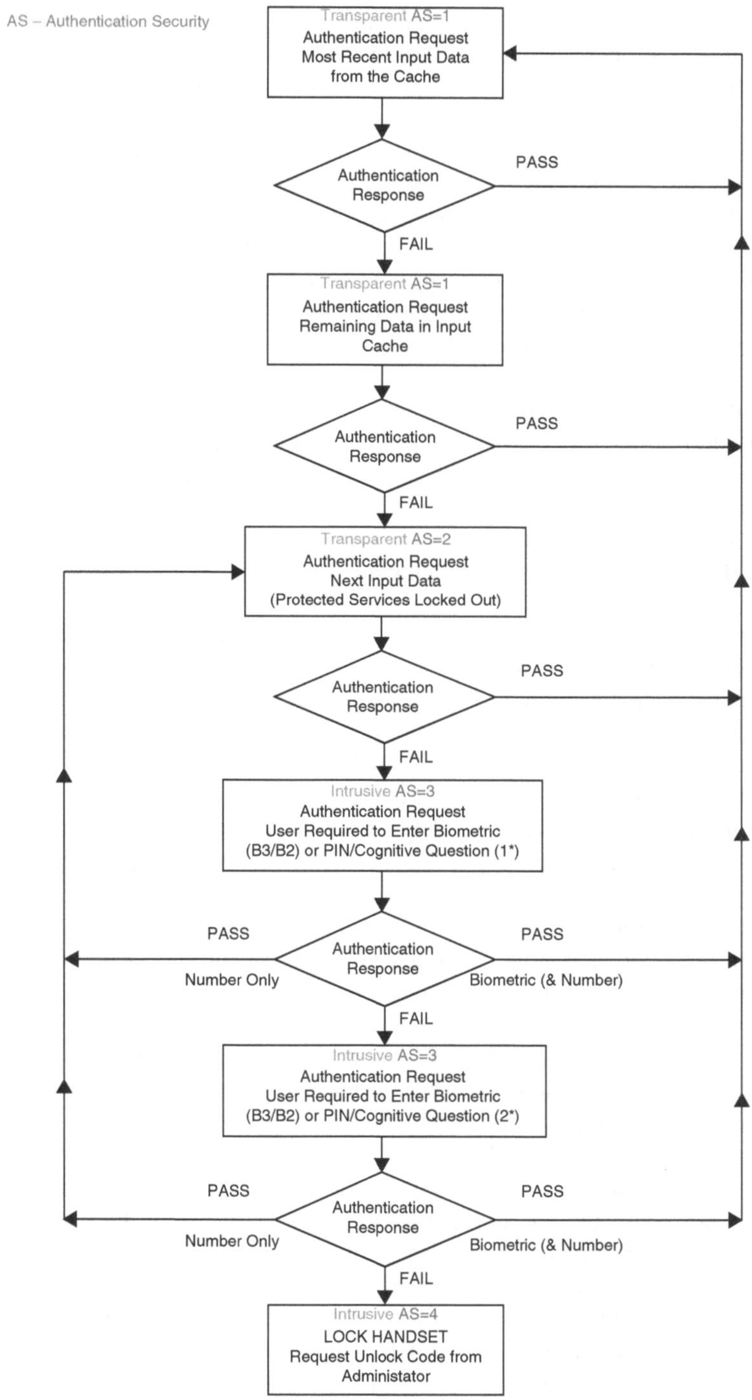

Fig. 8.13 NICA – authentication manager process

highest confidence level technique. If no additional data are present or the response is a fail, the Authentication Manager increases the security level and requests authentication on the next input sample to the device – the user would now not be able to use any of the protected services until this stage had been completed. If the user passes this, or any of the previous stages, then the Authentication Manager goes back into a monitoring/collection mode. If the Authentication Engine responds with a fail, then the Authentication Manager will request the user to authenticate himself or herself – the first intrusive authentication request. The Authentication Manager, via the Intrusion Interface, will use a biometric technique with a confidence value of B3 or B2 (i.e. if a system has both B2- and B3-type biometrics, then it will always use the higher of the two in order to minimise the risk of a false acceptance). If no biometric techniques or templates exist with a confidence value of B3 or B2, then the user will be requested to enter the PIN, password or answer a cognitive question. If the user passes this, and the PIN or password has a corresponding keystroke analysis template, then this will also be utilised in order to provide a two-factor authentication mechanism. If the keystroke analysis template exists, and the user passes the biometric authentication, then the system will revert to a monitoring mode. If the biometric fails or the template does not exist, then the technique will remain at a heightened security level of 2 – where the Authentication Manager will request authentication on the next available input sample. If an intrusive authentication request is passed, the previous biometric samples that were failed are deemed to be in fact from the authorised user and incorrectly failed. As such, these samples are added to the Profile database for subsequent retraining and are not deleted. This provides a mechanism for managing templates and template ageing – particularly for techniques that suffer from poorer degrees of permanence.

The Process algorithm is inherently biased towards the authorised user as he or she is given three non-intrusive chances to authenticate correctly, with two subsequent additional intrusive chances. This enables the system to minimise any user inconvenience from the authorised user perspective. However, due to the trade-off between the error rates, this has a detrimental effect on the FAR, increasing the probability of wrongfully accepting an impostor every time an authentication request is sent. With the Process algorithm in mind, for an impostor to be locked out of the device he or she must have the authentication request rejected a maximum of five consecutive times. However, this is where the System Integrity plays a significant role. The probability of an impostor continually being accepted by the framework becomes very small as the number of authentication requests increases. This would indicate that the impostor will be identified correctly more often that not (even if not consecutively as required by the Process Algorithm), reducing the System Integrity value to a level where the majority if not all of the services and file access permissions have been removed – essentially locking the device from any practical use. In a practical situation, it is likely an impostor will be able to make a telephone call or send a text message before the system locks down (the actual range of services available to the impostor will largely depend upon the Authentication Response table). However, all of the key sensitive and expensive services will be locked out of use. By permitting this limited time-constrained misuse of the device,

Table 8.6 NICA – Authentication performance

Mobile device	Authentication techniques	FRR at stage 4 of the process algorithm (%)	FAR at a system integrity level of +5 (%)
Nokia X3	Keystroke analysisvoice verification	0.001–0.4	0.000001–0.00002
Toshiba portege G500	Facial recognition Fingerprint scanning Voice verification	0.00003–0.0001	0.00000007–0.0000008
Apple iPhone	Facial recognition Keystroke analysis Voice verification	0.0002–0.4	0.0000008–0.00002

it is possible to achieve a much higher level of user convenience at minimal expense to the security.

The remaining duties of the Authentication Manager are mainly concerned with control, such as sending a template generation and retraining request to the Biometric Profile Engine (during computational non-intensive periods in a standalone topology) to update the biometric profiles and subsequently updating the Authentication Assets as a result. The client side Authentication Manager will also monitor (via the Communications Engine) the network connectivity to the server – should connection be lost at any stage, the client side will revert into a standalone configuration, thereby achieving autonomous operation. This could be useful in a number of situations, such as a poor network signal and network failure.

8.4.4 Performance Characteristics

The performance of such a composite authentication mechanism will be largely dependent upon the authentication techniques available to a particular mobile device. Those with stronger techniques will be more capable of successfully detecting an authorised and unauthorised user than their counterparts. Table 8.6 illustrates the performance achieved for a number of test cases based upon the authentication techniques that could potentially be available given their specific hardware configuration. As this composite mechanism involves multiple authentication requests and multiple authentication techniques it is difficult to obtain a single FAR and FRR. Table 8.6 presents the FRR at the point where the authorised user is essentially locked out from using the device, and the FAR of an unauthorised user achieving a System Integrity level of +5, which would permit the user to access the most sensitive services and information. The FAR and FRR for the authentication techniques which the subsequent system level performances were calculated from were derived from studies performed on keystroke analysis (Clarke and Furnell 2003) and the National Physical Laboratory (Mansfield et al. 2001).

Worked Example
FRR at Stage 4 of the Process Algorithm:

$$\begin{aligned}\text{Best case probability} &= \text{Voice FRR} \times \text{Voice FRR} \times \text{Voice FRR} \times \text{PIN FRR} \times \text{PIN FRR}\\ &= 0.04 \times 0.04 \times 0.04 \times 0.4 \times 0.4\\ &= 0.0000102\\ &= 0.001\%\\ \text{Worst case probability} &= \text{Tele FRR} \times \text{Tele FRR} \times \text{Tele FRR} \times \text{PIN FRR} \times \text{PIN FRR}\\ &= 0.29 \times 0.29 \times 0.29 \times 0.4 \times 0.4\\ &= 0.00039\\ &= 0.04\%\end{aligned}$$

Even with devices such as the mobile handset, with limited authentication techniques, the levels of FAR and FRR achieved are still stronger than many individual authentication techniques alone, with a (worst case) probability of an authorised user incorrectly being rejected of 0.4% (equivalent FRR) and a (worst case) probability of an unauthorised user gaining entry to the most sensitive services of 0.00002% (equivalent FAR).

The results from the theoretical system performance illustrate how difficult it is for an impostor to obtain access to sensitive services, with a FAR in the range of 0.00000007–0.000001% compared with the best FAR of 0.1% using a fingerprint technique. The false rejection probability has also improved, with a worst case of 0.4% and a best case of 0.00003%. Although it is difficult to directly compare the performance of this composite system against individual techniques (as the probability of successfully authenticating a person depends on various stages of the security algorithms), a comparison of these results against individual results illustrates the improvement in performance this mechanism experiences.

8.5 Summary

A TAS significantly increases the complexity of the underlying authentication mechanism. It is no longer merely a process of comparing two strings to see if they match but rather a complex process that has a number of permutations depending upon the technology platform and user interactions. Whilst the underlying process is complicated and the design requires considerable thought based upon the particular requirements needed, the actual function of the TAS is designed from the end-user perspective to be seamless – providing a more robust, reliable and continuous Identity Confidence, whilst at the same time reducing user inconvenience.

Implementation in practice, however, requires further consideration by the TAS designer. Aspects of privacy, storage, network connectivity, availability, mobility and cost all play a contributory role in whether a TAS is practically viable.

References

Clarke, N.L.: Advanced user authentication for mobile devices. PhD Thesis, University of Plymouth, Plymouth (2004)

Clarke, N.L., Furnell, S.M.: Keystroke dynamics on a mobile handset: a feasibility study. Inform. Manage. Comput. Secur. **11**(4), 161–166 (2003)

Clarke, N.L., Furnell, S.M.: A composite user authentication architecture for mobile devices. J. Inform. Warfare **5**(2), 11–29 (2006)

Karatzouni, S., Clarke, N.L., Furnell, S.M.: NICA design specification. University of Plymouth. Available at: http://www.cscan.org/nica/ (2007). Accessed 10 Apr 2011

Mansfield, T., Kelly, G., Chandler, D., Kane, J.: Biometric Product Testing: Final Report. Crown Copyright, UK (2001)

Monrose, F., Reiter, M., Wetzel, S.: Password hardening based upon keystroke dynamics. In: Proceedings of the 6th ACM Conference on Computer and Communications Security, New York (1999)

Chapter 9
Implementation Considerations in Ubiquitous Networks

9.1 Introduction

The implementation of a Transparent Authentication System (TAS) can take one of three forms: device-centric, network-centric or a hybrid combination of the two approaches. Each topology offers a different combination of enablers and restrictions. Even the hybrid model, which seeks to take advantage of both models, introduces its own issues – namely increased complexity. To better understand some of the practical issues that have to be considered in the implementation and deployment of a TAS, it is important to establish the specific set of user requirements. From these requirements, it is possible to determine which topology would provide the best fit. Key areas to establish the requirements and trade-offs that exist are:

- User privacy
- Storage, processing of biometric samples
- Network bandwidth requirements
- Mobility and network availability

The following sections address and discuss these issues, examining in detail the trade-offs between the two potential topologies.

9.2 Privacy

When considering which topology to deploy, resolving the issue of user privacy is essential for widespread adoption. This becomes even more important when the topology is looking to utilise biometric techniques as the underlying mechanism. Recent years have seen widespread media attention directed towards biometrics, due largely to their inclusion within passport and national identity card schemes (Gomm 2005). Unfortunately, and for some legitimate reasons, this attention has been somewhat negative towards the benefits of the technology, focusing instead

DOI 10.1007/978-0-85729-805-8_9, © Springer-Verlag London Limited 2011

upon privacy concerns (Porter 2004; TimesOnLine 2004). It is therefore important to ensure the authentication mechanism is designed in a fashion that is sensitive to privacy issues.

The principal concern focuses around the biometric template and sample. In whichever biometric technique that is used, these elements represent unique personal information. Unfortunately, unlike other forms of authentication (such as secret knowledge or tokens, which can be simply changed if lost or stolen), it is not possible (or necessarily easy) to change or replace biometric characteristics – they are an inherent part of the person. Therefore, once lost or stolen, they remain compromised and can no longer be reliably used. As such, the creation and storage of a biometric template or profile on either the device or the network leads to significant responsibility for the user or the network provider respectively.

Public opinion regarding biometrics has been problematic, not least because of the proposed national ID scheme in the UK. It called for a centralised repository of biometric information for UK nationals, but the ability to secure such databases from external attack and effectively manage authorisation to protect from internal misuse is no small undertaking. Despite the safeguards that one can apply, there will always be the potential for vulnerabilities due to both the human factors and technical mis-configurations. Such vulnerability, and moreover the lack of confidence that it engenders, was also raised in a focus group that took place in order to acquire users' views and attitude towards security on their mobile devices (Karatzouni et al. 2007), where participants voiced the concern over security and trust:

> [W]ould you really want your biometric data then stored on the inside of a company that's possibly got people dodgy, people breaking into it already.
>
> And even in the network don't think it's all that secure either, because there is always the rogue employee somewhere, who is in the pay of an attacker.

These quotes demonstrate a major fear for the security of the information held remotely. Apart from the technicalities that might be overlooked, there are also examples of carelessness taking place that has led to severe incidents. An illustrative example occurred within an Orange call centre, where employees who had been granted access to full customer records (including information such as bank details) were sharing their login credentials with other staff (Mobile Business 2006). This removed any ability to effectively monitor who had access to information and when. The increased fear of identity theft and fraud makes people even more cautious about their personal information, and how and where they provide it.

With the recent discussion of the UK ID scheme, it can be seen that people are not very comfortable with providing their biometric information to such a centralised system (Lettice 2006). As such a device-centric implementation is arguably more favourable from a user perspective. In such a case, the profile will be stored on their personal device giving no third-party access to the biometric template or samples. This approach is able to satisfy people's desire for privacy preservation through giving them direct responsibility for its protection. Nevertheless, such reliance does impose concerns about how reliable and also aware the end users will be in safeguarding their devices. As previously mentioned, several surveys have demonstrated that despite the storage of sensitive information in mobile handsets, and despite the earlier cited evidence of loss and theft, users still disregard the use of even the available security measures. This is an important consideration to the choice in topology, as no further protection will be available once the device is stolen. On the other hand, one might suggest that as the fear of misuse becomes greater, the importance that each subscriber will attribute to each device will change respectively. Storing personal identifiers in the device might lead people to consider their device to be comparable to other forms of important information and ID, such as passports, credit cards and car keys. Such linkage could potentially change people's perception and attitude towards the security and protection of their devices.

However, there was also a concern raised in the focus group expressing a fear of storage on the device and potential misuse:

> ... my concern is where would the fingerprint, let's say like signature, where would be stored? Would that be stored on the phone, so if somebody stole my phone they have my signature which is signed on the back of you bank cards and my fingerprint obviously? What then can people do with the information...obviously if someone knows how to hack into a phone could they use the information?

It is certain that a biometric database will always constitute an attractive target, making it a more valuable target than a device involving only one person. It would be necessary in such cases to establish regulations and policies for the security of the database and biometric templates, and mandate continuing adherence to them. A central system, though, has an advantage that the system can monitor such activity and try to prevent it, thereby providing a more uniform and controlled protection space than storage in the device.

People have different views towards the storage of such information as concerns are raised over the security in each storage solution and how potentially easy a breach of confidentiality is. A study conducted by the authors' research group attempted to assess public perceptions of biometrics, and performed a survey involving 209 respondents (Furnell and Evangelatos 2007). One question asked people about their concern regarding the theft of their biometric data and the potential of using them to cheat a system. The responses are illustrated in Fig. 9.1.

As seen from the figure the majority of the respondents expressed some level of concern about the security of their data, with only 4% not having any fear of misuse. The same survey also asked where respondents would prefer their biometric data to be stored. Forty percent supported the network option having the template stored in a central database whereas only 17% agreed on the device, as illustrated in Fig. 9.2. Interestingly, 18% would prefer their biometric templates to be stored in a smartcard. This is analogous to a device-centric approach, as the smartcard must remain with the user, but represents a significant enhancement in physical and logical security of the information.

Privacy concerns that exist for the network implementation could be reduced by ensuring only the biometric templates are stored and not any form of raw data. Several studies have taken place to overcome that issue looking to protect the storage of biometrics using techniques such as distortion of the template. It is also notable that

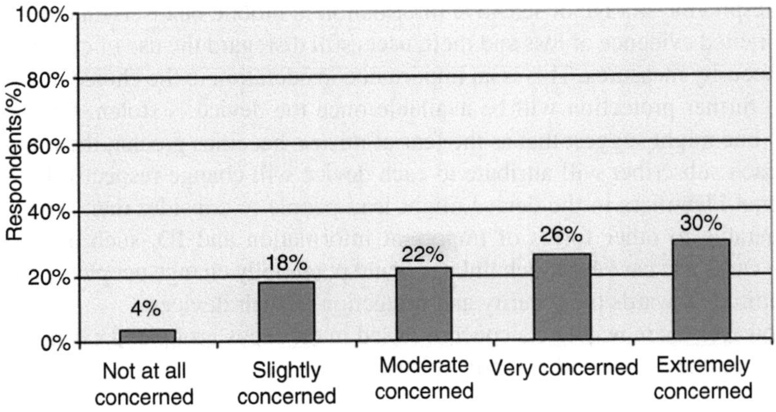

Fig. 9.1 Level of concern over theft of biometric information

Fig. 9.2 User preferences on location of biometric storage

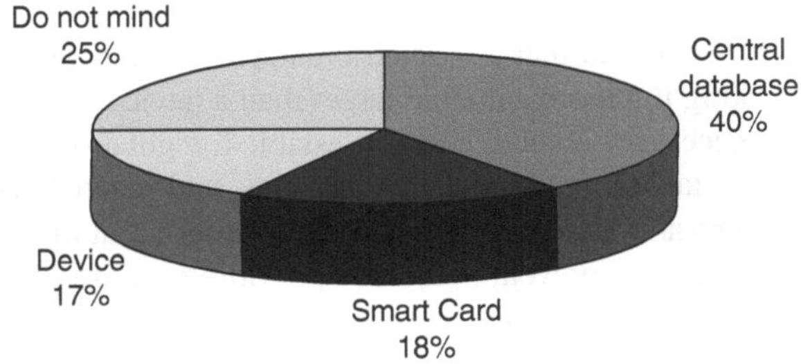

the creation of biometric templates is based upon vendors' own proprietary formats. As such, one biometric template from one vendor will not necessarily operate with another vendor's product, as the format and characteristics used to authenticate people differ. This will reduce the potential harm caused by a stolen biometric sample to systems that only utilised that specific vendor's product. The one-way property of creating biometric templates also ensures they cannot be reverse-engineered.

9.3 Storage and Processing Requirements

While the privacy issue represents a challenge of user trust and perception, there are also technical-level considerations in terms of the storage and processing of biometric data. These will again differ according to the chosen topology.

Consideration needs to be given to the storage of the initial biometric template and also the samples that are subsequently used in the process of verification. For current PIN-based approaches this is not an issue, but the storage demands of biometrics are more significant. Issues of storage might exist in both topologies, with individual

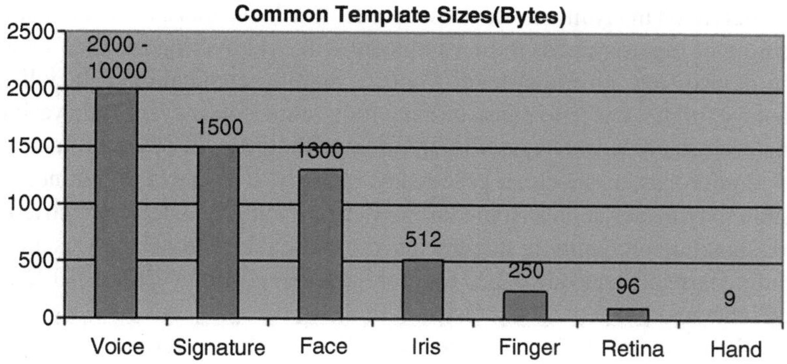

Fig. 9.3 Size of biometric templates

devices potentially having limited onboard storage, while the network-centric approach may need to cope with the storage of data for high volumes of users.

Different biometric techniques require differing levels of storage memory. Techniques such as face recognition (where multiple images might be needed from different angles in order to achieve a high consistent outcome) or voice verification (where sound files need to be stored) usually require higher storage capacities. Furthermore, as the proposed authentication mechanism aims to take advantage of a number of different techniques, the device or network will need to store more than one template per user, which could potentially become very demanding. Figure 9.3 illustrates typical template sizes from a number of more common biometric technologies.

Given the memory available on current mobile devices, it can be seen that the storage requirements would not prevent a device-centric implementation. The most demanding approach is voice scanning, which can reach the requirements close to 10 KB. Therefore, in general terms, storage of biometric templates in a device-centric model does not present any difficulty. However, given the variability in devices and functionality, some care must be taken to ensure that this proposed authentication mechanism is able to operate with all hardware devices, including legacy devices that might have smaller storage footprints. Moreover, the TAS approach allows for the capture and storage of all samples and, over time, this could result in a significant storage footprint. Appropriate maintenance of the storage, whether device or network-based, is required through policies to ensure a smooth operation.

In terms of processing capabilities, the network-centric approach has an advantage in the sense that devices themselves may have relatively limited capabilities. Indeed, this may actually represent a fundamental obstacle to establishing a device-centric solution. While laptop-level devices may have the capabilities required to process biometric data, the processing power in handheld devices is still limited. Algorithms that are utilised in biometric verification tend to be intensive, as they are based upon complex data extraction and pattern classification techniques (and indeed the impact of this additional processing on the battery of the mobile device would also have to be carefully considered). For desktop-based systems this is obviously less relevant. The process of enrolment and verification will place a serious demand upon resources on many mobile devices. In order to achieve transparent authentication, verification of the user needs to be completed without affecting the user's ability to use the device (e.g. no impairment to other running applications). It will not be satisfactory for the device to pause or hang for a few seconds every time verification is being performed. However, as with the storage footprint, different biometric techniques require varying levels of processing capacity. It is therefore not necessarily infeasible to consider at least some biometrics operating in a device-centric model. Indeed, signature recognition, fingerprint recognition, keystroke analysis and facial recognition have all been developed for mobile devices (Turner 2001; NTT DoCoMo 2003; Clarke and Furnell 2006a; Omron 2005).

Over time, the processing constraints are likely to be overcome in the future as the capabilities of handheld devices continue to advance. However, from an implementation perspective, a network-centric model would still potentially be easier to deploy and offer a wider range of possible biometric techniques that could be used. Again, however, consideration needs to be given to the scalability of such an approach – multiplying individual authentication requests by high volumes of users does place a significant demand upon processing.

9.4 Bandwidth Requirements

A particular consideration in the context of the network-centric approach is the network bandwidth that will be required for the transmission of user authentication data. A device-centric approach has no such implications, as at most it will only be required to perform its normal authentication of the device to the network. By contrast, the network-centric approach will require network bandwidth to send biometric samples to the network, and receive authentication decisions back.

Whether the TAS is deployed on a device with wired or wireless connectivity, the network-centric model will introduce a delay in the authentication process. This willingness to wait is an important consideration in designing the authentication protocols and mechanisms. Forcing users to wait too long before being given access would result in a negative perception, particularly when the approach is meant to be user-convenient. It is therefore important to consider the networked environment within which the TAS is going to be deployed and ensure sufficient network bandwidth is available. For transparent authentication, the network latency is less of an issue. However, for intrusive authentication, users would expect a timely response.

As previously discussed, biometric templates can range from as little as a few 100 bytes up to 10 Kb. These templates contain the unique data that are derived after pre-processing (thereby extracting required features). The option of the device performing this procedure would be a way to decrease the bandwidth requirements as the data sent would be far smaller than the raw sample. However, the ability to perform pre-processing on the device will depend upon the individual biometric technique and the processing capabilities of the device. If pre-processing can be implemented on the device it can be assumed that the size of data being communicated is

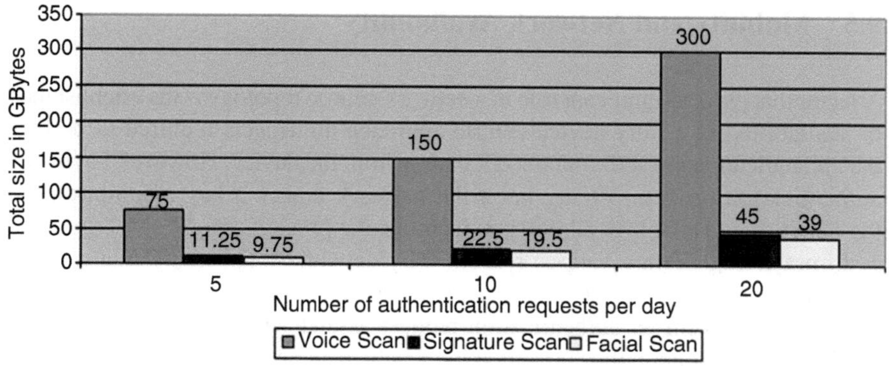

Fig. 9.4 Average biometric data transfer requirements (based upon 1.5 million users)

similar to those presented in Fig. 9.3. A simple computation will show that the largest template of 10 Kb will require a time of 0.36 s for the lowest given throughput of UTMS (220 kbps). It must be considered though that this might well become larger depending upon the network condition at the time and also takes no consideration of the time taken for the network to actually perform the authentication. Beyond latency for individual users, the issue of scalability needs to be addressed. Large volumes of users sending biometric sample data across the network might have a significant impact upon network resources and increase the level of delay experienced. For example, if a mobile network operator were to deploy a TAS, this could involve well over 15 million subscribers. If only 10% of them used such a service, it would involve 1.5 million users requesting authentication from the network. Of course the burden on the network will depend upon the authentication frequency and this will vary across users, as the different use of their device will result in more or less authentication requests. Nevertheless, the scalability of the TAS ramps up considerably when dealing with a large population of users.

At first glance, one might suggest that current 3G networks (and certainly future networks) would be able to cope with the requirements. Although this might not be wrong in principle, an investigation of the network consumption does reveal somewhat surprisingly high volumes. Based upon the figure of 1.5 million users, Fig. 9.4 illustrates the bandwidth required per day for three different types of biometric approach.

As illustrated in the chart, for voice scan, even a minimum request for authentication of five times results in a required data capacity of 75 GB for the network provider, whereas with up to 20 requests per day this rises to 300 GB. In comparison, however, a video stream application (one of the standard mobile phone applications) has bandwidth consumption close to 200 Kbps for each user. In a population of 1.5 million subscribers that represents 37 GB to be transferred every second. That said, there is a real cost associated with sending data across a network and there will be at least an indirect cost, given that the operator may otherwise be able to use the bandwidth to support revenue-generating services.

9.5 Mobility and Network Availability

A factor that plays a significant role in a network-centric topology is the establishment of availability. In a fully device-centric approach all aspects required to perform the authentication are self-contained locally within the device. However, having the authentication process relying upon the network makes a key assumption that the network is available at all times to facilitate the process.

In practice, there are various reasons why network connectivity might not be available, such as loss of coverage, network overload, or server malfunction. The inability to perform an authentication request as and when required will have a significant impact upon the authentication mechanism and its perceived usability.

Of course, if the authentication request is associated with a network-based application or service then one could reasonably argue that there is no inconvenience, as the service would not be available anyway. What would be less acceptable, however, would be a reliance upon network availability in order to access applications or features that would otherwise be entirely local. For example, opening a document, accessing contacts or using Bluetooth to connect to another device might all require authentication, and this would have a real and unacceptable impact if the process were to rely upon the (unavailable) network.

Participants in the focus group were asked to consider this issue and overall there was a negative opinion on always requiring access from the network. The following viewpoint was typical:

> I find it difficult that it might be possible just even to interact with the network operator, because I'd like to use that information even when I don't interact with the network operator.

It can be suggested that apart from the technical issues that can occur, it seems rather inconvenient to require authentication from the provider. The inconvenience does not only relate to the access of local functionality and applications but also to the general concept that in order to access any service the user will be obliged to explicitly go through the network provider. This places a burden of inconvenience upon the user, network provider and the authentication mechanism. One of the focus group members specifically summed up the issues surrounding the availability of network resources:

> There is quite a lot stored in the network. Potentially everything can be stored in the network. There is a trade-off between responsiveness and security. …Especially if you are not in coverage all period of time and you want to look up someone's name, address or whatever in your address book you haven't got it. So that's completely rejected by the operators. There's got to be some balance between security that happens on the network and immediacy you have on the person … there isn't a simple answer to this sort of question.

This issue is further exacerbated when considering the issue of roaming. A network-centric topology would experience significant increases in latency and have to transverse a far larger open network. Unless the local network provider supported the authentication mechanism and had a local version of the biometric template (which would not be likely due to privacy) this increase in delay would again have an impact upon the authentication mechanism which would need to be considered. The device-centric topology is more appropriate as authentication of the user can be performed on the device wherever they might be in the world.

But what also happens when roaming is not available? In such a situation, the user will have no way to be authenticated as no access to the provider's network will be available, restricting if not completely preventing any use of the device. There is also the consideration of cost. A home network operator implementing the authentication mechanism might be prepared to bear the cost of network consumption. However, this may not be the case for a roaming network, raising questions of who covers the cost. A device-centric approach would overcome this issue as no reliance upon external resources is required.

The prior analysis has shown that comparing the device- and network-centric topologies introduces a varied and complex range of considerations, with each approach offering advantages and disadvantages in different contexts. Attempting to base a solution entirely around the device can introduce processing limitations, whereas bandwidth and the requirement for connectivity may represent practical constraints for a network-based model. In addition, both approaches may introduce their own privacy-related concerns.

9.6 Summary

Based on the issues arising from both principal architectures (device and network), it can be seen that no single approach can cover all requirements for practical implementation of a TAS. Each TAS environment needs to be carefully analysed to understand and appreciate the relevant trade-offs that exist and develop a model that best fits the requirements. In many situations this would result in hybrid-based systems.

In such an approach both the storage and processing could be potentially split over the device and the network, compromising between the issues of device-processing capabilities, network availability and privacy. The nature of the split in the authentication mechanism will depend upon the individual requirements of the user/system in relation to privacy and access, and the device in terms of which biometric techniques it can support locally. There will be therefore a number of hybrid approaches that could exist, each covering different issues on different scenarios for different users. For example, in order to deal with the issue of device processing and privacy, there could be the option to store all of the templates in the device, but place the processing functionality on the network. This would satisfy privacy concerns but at the same time discharge the device of any excessive processing tasks. Cryptographic measures could be used to protect the data in transit and during processing. Depending upon the device capabilities, pre-processing can be performed locally when possible, so that the biometric samples that are being sent over the network are kept as small as possible. The specific nature of the hybrid system will closely depend upon a wide variety of factors. In order to remove the concerns surrounding network availability, it is suggested that at least one authentication technique always remains on the local device. Although this technique might not provide the level of security strong network-based biometrics might, it will be able to provide an effective means of authenticating short-term usage of local applications and functions during periods of network inaccessibility. In devices with more processing capacity, the hybrid approach would also be able to provide the ability to split the biometric templates, having the most intensive and demanding biometric techniques on the network and the others with fewer requirements on the device. Another basis for determining this split could also be the uniqueness attributed to them for privacy issues.

Acknowledgement Aspects of this chapter have been taken with permission from Karatzouni S., Clarke, N.L, Furnell, S.M. (2007) "Device- versus Nework- Centric Authentication Paradigms for Mobile Devices: Opertional and Perceptual Trade-Off's". Proceedings of the 5th Australian Information Security Management Conference.

References

Clarke, N.L., Furnell, S.M.: Authenticating mobile phone users using keystroke analysis. Int. J. Inf. Secur. **6**(1), 1–14 (2006a)

Clarke, N.L., Furnell, S.M.: A composite user authentication architecture for mobile devices. J. Inf. Warfare **5**(2), 11–29 (2006b)

Furnell, S., Evangelatos, K.: Public awareness and perceptions of biometrics. Comput. Fraud Secur. **2007**(1), 8–13 (2007)

Gomm, K.: Full biometric ID scheme to reach the UK 'by 2009. ZDNet. Available at: http://news.zdnet.co.uk/hardware/0,1000000091,39232692,00.htm (2005). Accessed 10 Apr 2011

Karatzouni, S., Furnell, S.M., Clarke, N.L., Botha, R.A.: Perceptions of user authentication on mobile devices. In: Proceedings of the ISOneWorld Conference, Las Vegas (2007)

Lettice, J.: Compulsory and centralised – UK picks hardest sell for ID cards. The Register. Available at: http://www.theregister.co.uk/2006/03/13/ou_idcard_study/ (2006). Accessed 10 Apr 2011

Mobile Business: Orange data not secure. MB Magazine. Available at: http://www.mbmagazine.co.uk/index.php?option=com_content&task=view&id=1441&Itemid=2&PHPSESSID=d6fc7ad0c429dae5956c3ffd9466a84d (2006). Accessed 10 Apr 2011

NTT DoCoMo: NTT DoCoMo unveils ultimate 3G i-mode phones: FOMA 900i series. NTT DoCoMo. Available at: http://www.nttdocomo.com/pr/2003/001130.html (2003). Accessed 10 Apr 2011

Omron: Omron announces "OKAO vision face recognition sensor". World's first face recognition technology for mobile phones. Omron. Available at: http://www.omron.com/news/n_280205.html (2005). Accessed 10 Apr 2011

Porter, H.: If you value your freedom, reject this sinister ID card. Guardian. Available at: http://www.guardian.co.uk/idcards/story/0,15642,1375858,00.html (2004). Accessed 10 Apr 2011

TimesOnLine: ID and ego: it is right to experiment with identity cards. The Times. Available at: http://www.timesonline.co.uk/article/0,,542-1089392,00.html (2004). Accessed 10 Apr 2011

Turner, I.: Eye Net Watch launch PDALok for IPAQ. computing.Net. Available at: http://www.computing.net/answers/pda/eye-net-watch-launch-pdalok-for-ipaq/208.html (2001). Accessed 10 Apr 2011

Chapter 10
Evolving Technology and the Future for Authentication

10.1 Introduction

The need for changes in the way authentication is performed has been clearly identified in the early chapters. Indeed the purpose of this text is to provide a different perspective on the problem – moving away from the traditional point-of-entry Boolean decision to a continuous identity confidence. The Transparent Authentication System (TAS) seeks to harness all authentication approaches and provide a level of flexibility to enable compatibility and acceptance from all aspects of society, the young, old, professional, unskilled and security-indifferent individual. However, this approach is far from being a panacea for the authentication problem. It introduces a level of complexity into the authentication process that goes far beyond the provision of a username and password. The ability of a TAS to succeed will depend upon its ability to adapt:

- Adapt to better understand the authentication requirements of a system and the individual – to provide commensurate protection for the task at hand.
- Adapt to the evolving technological requirements, new technology platforms and new interface technologies to ensure authentication is being performed in the most convenient and user-friendly fashion.
- Adapt to consider not only the individual authentication requirements of a system but also the individual authentication requirements of the person. Through disassociating the process of authentication with individual systems and associating it with people, authentication between systems can be cooperative, further enhancing the user experience.

The following sections in this chapter will look at each of these aspects in turn: understanding what future developments would improve the efficiency of a TAS through the implementation of more intelligent and adaptive systems, extending the technological trends and considering what and how the authentication requirements might change as technology evolves and, finally, appreciating what the next step in authentication technology might be.

N. Clarke, *Transparent User Authentication: Biometrics, RFID and Behavioural Profiling*, 215
DOI 10.1007/978-0-85729-805-8_10, © Springer-Verlag London Limited 2011

10.2 Intelligent and Adaptive Systems

Intelligence is a word frequently used by security practitioners. Systems that can adapt, evolve and learn are an incredibly useful attribute where the requirements are continuously changing. For instance, in security, the evolving threat landscape provides a dynamic environment from which security countermeasures must adapt to ensure systems remain secure.

TAS already advances the concept of authentication beyond a point-of-entry, one-off, authenticate-and-forget strategy to an approach that considers the needs of both the system and the individual, providing a mechanism by which the system has a continuous appreciation of the current user. However, interactions and service requests – opening a document or running an application – is still considered on a case-by-case basis. Should a user have sufficient identity confidence, access is granted. If not, further verification of a user's identity would be required. TAS is not able to appreciate or understand the relationship between interactions. To understand that whilst individual access to several files might only require a low level of confidence, they might indeed collectively reveal a far higher degree of sensitive information. If stored in a single file it may require a higher level of confidence in the user's identity. For example, the running of a tool such as Netcat to understand what network connections and ports are open might not in itself appear to require a high level of identity confidence. However, if the user had also run other applications, such as listing the applications and their versions, or using File explorer to search around the file system, collectively a high level of concern would exist that the current user is undertaking some level of reconnaissance on the system. Each individual action is benign, but together they tell a completely different story.

An improvement in a TAS would be for the system not to be constrained to a particular authentication decision but have a higher-level situational awareness of the system, the access requests that have been made, the services that are currently running and an understanding of what the user might be trying to achieve. Through providing a more holistic approach to the management of the authentication system, more appropriate and intelligent decisions can be made. In this manner, particularly for less sensitive material, the process of access control verification can move away from individual identities towards a process-orientated access control. In addition to being able to make more informed access control decisions, it will also reduce the number of access control requests, reducing the possible burden upon end-users.

The ability for a TAS to adapt and respond intelligently in this manner will also help in minimising the problem of attackers sweeping up the low-level information. A design feature of a TAS is to effectively trade off user inconvenience and security. Through understanding the risk associated to particular access requests, the system is able to ensure a commensurate level of security. The negative aspect to this feature is it opens a window of opportunity for an attacker to obtain access to information – particularly the data that has an associated low level of risk. Whilst the level of access and window of time available are dependent upon the configuration options and the state of the identity confidence at that specific point in time, it can in certain situations provide an attacker with sufficient indirect information to enable a further attack.

The requirement for intelligence does not just reside with the TAS but also with the underlying biometric techniques. Ignoring their application specifically in a transparent fashion, intrusive biometric techniques also suffer from a variety of issues, including usability, acceptance and performance. A key issue is also the highly constrained operational environment within which they need to operate in order to perform well. All techniques suffer from some form of environmental or physical challenge. Facial recognition suffers from problems in facial orientation and illumination, speaker recognition from background noise, keystroke analysis and behavioural profiling from short-term physical disabilities or issues (e.g. consumption of alcohol is likely to vary these characteristics) and fingerprint recognition from place, orientation and swiping speed. To apply these biometric techniques in a transparent fashion only serves to exacerbate the issue.

In many situations, algorithms exist that can perform extraction and classification of samples in various environmental conditions. However, in order to maximise performance, biometric vendors tend to tailor their algorithms to meet a particular set of requirements. Multi-algorithmic approaches might well prove to be the solution of providing additional robustness. The ability to include multiple algorithms provides a mechanism for the system to select the most appropriate algorithm given the scenario. For instance, if the orientation of the face is a frontal zero-degree orientation, the best-performing standard facial recognition algorithm can be used. Should the image contain a face at an angle, recognition algorithms optimised to extract and classify on such images would be utilised. In such a manner, a single biometric approach can appear to operate within a far larger set of operational characteristics. The work undertaken by the ISO/IEC JTC1 SC37 standard committee lays the foundation of this type of interoperability.

Furthermore, more intelligence needs to be developed when dealing with the spoofing of biometrics. The performance of biometrics tends to be always reported in terms of the system performance over a population of users. It fails to illustrate the necessary granularity of where the errors are coming from and does not truly reflect the problem. A false acceptance rate (FAR) of 3% does not mean each impostor will be accepted 3 out of every 100 attempts, as it might suggest. In practice what this means is that, whilst a good proportion of the impostor population will experience FAR much lower if not 0%, a small proportion of users will have a FAR much higher. The reason for this is that those users' samples happen to more closely match that of the authorised user – and so are accepted on a more frequent basis. Doddington et al. (1998) characterise users as follows:

- Sheep – a standard user
- Goat – a user who is particularly difficult to recognise
- Lamb – a user who is particularly easy to imitate and
- Wolf – a user who is able to imitate other users

Of particular interest are the lambs and wolves. Lambs present a problem to the system administrator in that whilst they might expect a particular level of performance from a biometric system (and indeed might have purchased the system on the understanding of how it can perform), lamb users will experience a far lower level of security.

In a practical scenario, identifying lambs is not a simple task unless the administrator runs a series of experiments against the templates. It is likely that a mixture of the two approaches will need to be taken. Within the biometric technique, measures are required that can more intelligently understand the nature of the feature space and how users' individual samples map onto this. In this manner, the technique could include a mechanism for recursively adding additional features (if available) until the feature vector is distinctive enough. This could be possible as many biometric techniques employ feature reduction algorithms to reduce the problem caused by the curse of dimensionality. Alternatively, the technique needs to inform the user or system that this user is likely to be an easier target. The second approach deals with how the system regards the biometric technique itself. Rather than basing confidence on the recognition performance of the technique as a whole, the recognition performance of the technique should be based upon the individual and what they explicitly achieve.

Wolves also represent a significant issue for the future success of biometric systems. Obviously, the use of a TAS utilising a range of authentication approaches does increase the challenge for an attacker spoofing the biometric. Rather than merely spoofing a sample when requested by the system, the attacker has to continuously spoof the samples, as they will have no knowledge of which will be used in the authentication process. Multiplying this across a range of biometric techniques would require continuous spoofing of all biometrics, which is not a simple task.

The more fundamental problem, however, is the fact that biometric techniques suffer from issues of spoofing in the first instance. Significant research has already focused upon mechanisms to reduce spoofing and, in particular, liveness detection. Intelligent mechanisms have already been devised to assess the sample and determine whether it has come from the legitimate person or from an artificial source. However, for every new approach devised to test liveness, attackers find a way of circumventing it (to one degree or another). Therefore, further intelligence and adaptability need to be incorporated within the biometric system in order to continuously counter the threat posed by spoofing.

Adding additional layers of information to the system will enable a TAS to act in a more intelligent fashion, making more informed decisions. Research in some of the underlying technologies such as computational intelligence has progressed significantly and is enabling a need breed of intelligent systems that adapt and respond as the environment evolves. Indeed, neural network approaches in particular are already widely used within biometric matching subsystems as more robust classifiers than their statistical counterparts.

10.3 Next-Generation Technology

It is difficult to predict exactly how technology will evolve. There are, however, a number of trends in both the underlying technology and within authentication that can be identified and extrapolated. Within technology the major trends are:

- Mobility of technology
- High degree of technology penetration

- Ease of availability and accessibility of information
- Blurring of boundaries between organisational and private systems and information

Mobility and penetration of technology is a relatively easy trend to evidence. The advancements in computing technology have resulted in portable computing devices that have an almost ubiquitous acceptance amongst people. In Western societies, mobile phone penetration based upon subscribers alone accounts for over a 100% of the population. This doesn't include users with laptops, PDAs, netbooks and tablets. In the last quarter of 2010 alone, Apple sold over 16.24 million iPhones and 7.33 million iPads (Apple 2011). This phenomenal demand for technology appears never-ending. The need for data anywhere at anytime is only going to increase, with people wanting a wider range of access to all nature of information, including, Web, music and video content. The demand for such data consumption can also be evidenced from the growing range of wireless-based data networks and specifically the increasing bandwidths they can support. The Global System for Mobile Communications (GSM)-based networks, once designed for telephony traffic only, can now support data rates up to 58 Mbit/s using High-Speed Packet Access (HSPA) over a 3G network, providing wide area coverage and high data communication bandwidth. Furthermore, what is being connected to the Internet is also on the increase. Anything that has any form of control associated to it is getting an Internet Protocol (IP) address – computers, fridges, lamps, central heating systems, cars, CCTV and industrial equipment. Technology is penetrating all aspects of society.

The desire for accessing information from anywhere at anytime has also created an expectation that such information will always be present. Moreover, any attempts to restrict access or availability of information will result in a high level of user annoyance. This demand will only increase. Environments that currently are unable to provide users with data will become evermore connected. Even airplanes, amongst the few places where Internet connectivity was not possible, are now equipped with wireless connectivity. Access and availability of information will become truly ubiquitous. Indeed, an increasing belief amongst people is that access to the Internet is a fundamental human right (BBC 2010).

There is also a trend that is blurring the boundaries between organisational and personal information. The ubiquitous nature of computing tends to result in people using whatever technology would most efficiently complete the task at hand. This means users will use their personal mobile phone to respond to a business email, use their home computer whilst at home to catch up on work, or use their work computer to do some online shopping or personal banking. Technology has become so ubiquitous that little consideration is given to what the device should be used for, merely whether it can achieve the desired aim.

These trends are all likely to continue as computing technology becomes more pervasive. Research and conceptual prototypes already exist of systems such as wearable computing devices woven into the fabric of clothing; bioinformatic sensors that measure and monitor our heart, blood pressure and medical conditions, and provide users and doctors with immediate notification of any health issues and heads-up display of information being provided in sunglasses – or even directly interfaced with our brains. These systems will all become part of everyday life as much as making a telephone call is today.

Each of these trends also introduces problems with regard to providing effective security. As the aforementioned trends push technology forward, the burden placed upon the user to maintain security will arguably significantly increase. But will the user take responsibility or indeed be in a position to do so. As more services are delivered via the Internet rather than traditional physical interactions (such as posting a letter, going to a bank to do your banking or accessing government services through your local office), all individuals across society will be forced to assimilate and use technology. With so much reliance upon technology to service our everyday needs, the impact upon service disruption will simply be unacceptable – both to the individual and to the nation as a whole. Security is therefore imperative for the future. The question is what role the user will play in maintaining security. History has demonstrated that technological solutions alone cannot solve the security problem. It is also not feasible to simply assume another stakeholder will provide security for you – particularly in a networked world where so many stakeholders exist, each looking out for the security of their particular area (and infrequently your data). Organisations will increasingly lose oversight of their information as it becomes increasingly distributed amongst employees' personal devices. It is inevitable that end-users will have to have more informed security practice and a better understanding of how to remain safe and secure. However, if it is going to be acceptable to end-users, it must not affect immediacy to information and must be provided in a fashion that all aspects of society will understand.

Authentication is core in providing user security. This requires authentication approaches that are reliable, convenient and applicable across platforms and services. With such frequent access to data, it will simply be infeasible for services to request intrusive authentication of the user. Single sign-on and Identity Federation schemes are already deployed to reduce this problem of multiple authentications and these are likely to mature and become more commonplace. They do not, however, solve the authentication problem. Once authenticated, the authorised user is assumed to be the person accessing the service. Transparent and continuous systems will have to provide the necessary confidence in authenticity.

Given the overhead and technical complexities of deploying a TAS, it is envisaged that the process of performing the authentication could be passed on to a managed authentication service – a dedicated company equipped with the necessary vendor-supplied biometric algorithms to provide robust identity verification in a timely manner. Much like how many managed security providers currently operate, managed authentication providers will provide a specialist service that enables more accurate authentication and monitoring of the user.

Inevitably new authentication techniques themselves will be devised. Research into DNA, brainwave and odour techniques are already under way. DNA and brainwave approaches offer the opportunity for highly reliable identity verification. Odour could potentially present a very transparent approach. Radio frequency identification (RFID)-based approaches have also evolved, being injected beneath the skin. Such approaches avoid the problem of forgetting to take the token or indeed theft of the token. However, the injection of the RFID chip underneath the skin does arguably present a somewhat overly intrusive step to providing effective authentication. The extent to which these new approaches will become effective is yet to be proven both in the laboratory and practically. However, it is safe to argue that none will by itself provide the panacea to the authentication. None will be ideal for all users, for all devices and in all situations. The future for authentication has to be directly connected to enabling users – not inhibiting them. Therefore, intelligent implementation of multibiometrics and composite authentication is the only future for effective, reliable and robust authentication of users.

10.4 Authentication Aura

Whilst transparent and continuous authentication of users helps to reduce the burden upon the user, it does so in a device-specific manner. If all your computing platforms support a TAS, then it will be able to non-intrusively verify a person's identity on a continual basis ensuring security is maximised and user inconvenience minimised. But each device in turn needs to be able to establish a level of confidence in the individual through the authentication techniques available to it. Whilst in some scenarios this is perfectly acceptable – and indeed the preferred and only option available – scenarios exist that suggest a TAS is overcomplicating the problem, with the possible effect of increasing user inconvenience. To explain let us begin with an example scenario:

> In the morning over breakfast you check your iPad for any email that has come in overnight and your mobile phone for any text or voice messages. You then continue to use your iPad to read the morning news and catch up on world events. On the train into work, you use your mobile phone to finish reading your favourite book. Upon arrival in the office, you sit down at your desk and start working through the email on your desktop computer.

The scenario is designed to illustrate a typical beginning to a working day for many people. In this scenario, the individual interfaces with a range of personal technologies on a number of occasions in a relatively short period – iPad, mobile phone and desktop computer. In each case, the TAS system will try to authenticate the individual transparently but as he or she has only just begun to use the devices, it is likely the system will not have sufficient confidence to provide immediate access. Instead, the user has to provide authentication intrusively, providing at least three sets of authentication credentials to each device, ignoring the additional credentials required for accessing the services. A significant amount of time can be spent logging and signing into systems and applications. Whilst TAS will reduce this burden as more interactions with the device occur, it can only do so whilst the user is interacting with that specific device.

However, if the device could leverage the authentication confidence from other devices, this would reduce the need for each individual device to capture samples. Instead, the confidence created by another device during user interactions is communicated to other devices that belong to the user. Using our previous example, when the user begins using the iPad to check emails, he or she will be authenticated. The result of that authentication (success or not) can be communicated to other devices that belong to the user. In this case, if the authentication request on the iPad was a success, it would provide immediate access to the mobile device – even though the mobile has no direct authentication samples to verifier identity. Similarly, as the user walks into the office, the fact he or she had been using the mobile phone on the journey would result in the device having established a good level of confidence in the user. When in close proximity to the desktop computer, it can communicate this to the personal computer (PC) and enable the user to be automatically logged in.

The approach disassociates authentication purely on an individual device basis and enables cooperative and distribution authentication between devices. The initial research was published by Hocking et al. (2010) and is referred to as an *Authentication Aura*. The concept is to remove authentication confidence from the individual device and associate it with the person, enabling him or her to enter and leave environments with the technology surrounding him or her, understanding who they are and providing access based upon the identity confidence.

In practice, a person cannot simply provide an aura. But the technology with which he or she interfaces and carries can. Every occasion a user interfaces with technology, the underlying TAS can capture samples and provide identity verification of the user. That confidence can also be broadcast out to all associated and trusted devices within a fixed proximity, providing an input to each of the other TASs, enabling them to refine their own locally defined identity confidence based upon the most recent authentication on the other device. In such a manner, each device still has responsibility of managing and providing the local decisions on access, but authentication samples processed elsewhere can also be used as an input into the system – almost as if the sample and authentication were completed locally. Figure 10.1 illustrates the cooperative and distributed nature of the authentication aura.

The advantage of such an approach is it can provide a more robust measure of confidence because it will be able to support a wider range of core authentication techniques than perhaps any single device can. So a device that can only support a couple of authentication techniques due to input and processing constraints can utilise the aura, which itself is based upon authentication decisions from other devices supporting a wider range of authentication techniques, to establish a stronger confidence in the identity of the user.

There are of course some very real security concerns with such an approach – the ability to trust another device to provide an input into the local authentication confidence introduces a number of threats regarding trust and attacks manipulating particular devices. For instance, the ability for an attacker to overcome one device would, because of the trusted relationship between devices, present a real threat to the remaining devices within the trusted network. However, this concept of trust between computing systems is not new, and mechanisms exist to minimise this threat. Moreover, the close physical proximity required to facilitate this process reduces the threat considerably. Other threats such as capturing network traffic and replay attacks are potential concerns that need to be considered when designing the protocol utilised to communicate but again protocols in the traditional network security domain offer solutions to these.

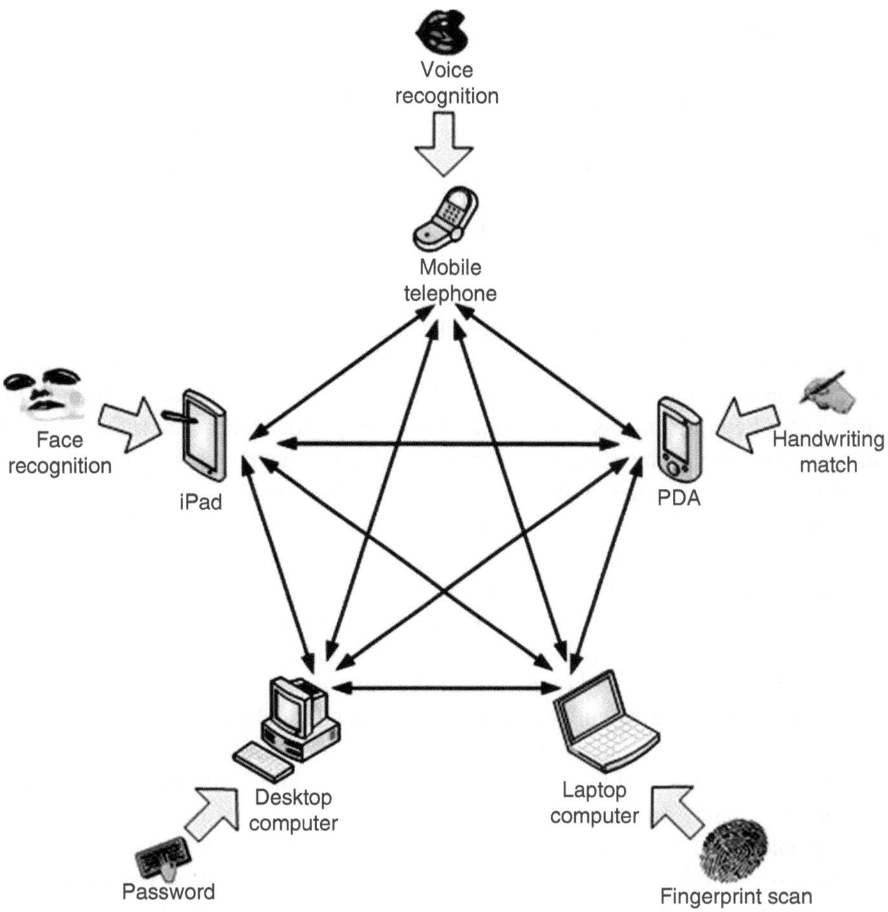

Fig. 10.1 Conceptual model of the authentication aura

The largest threats come as a result of the actual operation of the aura. How quickly the aura is created and more particularly dissipates can provide a window of opportunity to the attacker. Future research is currently being conducted looking to the nature of the interactions between devices in order to formally understand how these interactions can assist in establishing and reducing the authentication aura.

As the number of technologies increases and becomes fully integrated into the fabric of society, the increasing need to authenticate can be offset by the increased range of possible interactions between technology platforms. Whilst obvious security concerns would exist today, it can be envisaged that such techniques could also be utilised to authenticate third-party systems as well as personal devices – the ability to walk up to a cash machine and for it to understand who you are with sufficient confidence that access to services and withdrawing cash could be automatic. Such an approach offers a truly transparent mechanism for the authentication of users in a technology-driven society.

10.5 Summary

Information systems have undoubtedly enabled all aspects of society in the way we communicate, the way we work and what we do with our leisure time. As new technology is created, people become increasingly reliant upon it. Simple technologies such as email, text messaging and telephones provided services people did not think were even initially necessary. Yet today, we live in a society where the removal of such services is considered a breach of our fundamental human rights.

Information Security, once not considered an issue in the utopian concept of the Internet, has grown to become a significant barrier to the ubiquitous adoption of information technology. As the value of our interactions increases on the Internet and our reliance upon technology grows, so does the opportunity for cybercrime. Significant investment by governments and organisations looking to protect our critical national infrastructure is a prime example of the potential threat that could be faced.

Whilst good information security involves a holistic appreciation of all facets of security – technical, physical, procedural, human and legislative – the basic process of correctly authenticating the legitimate user is paramount to enabling secure systems. Authentication offers the attacker an easy way into systems – no need to identify software vulnerabilities through laborious iterative testing of applications, no need to craft network packets to circumvent sophisticated intrusion detection system (IDSs), no need to circumvent access control policies or seek root permission. It is imperative that authentication is robust and reliable.

The TAS and associated concepts introduced in this book have provided an insight into how authentication can be perceived differently, essentially identifying the point-of-entry one-off authentication mechanism as not fit for the purpose. To authenticate a user once and then provide continual access to the system and services is a significant security breach. To enable authentication on a continual basis forces designers to consider the issue of user inconvenience. More intelligent and user-convenient authentication approaches are essential.

References

Apple: Apple reports first quarter results. Apple Corporation. Available at: http://www.apple.com/uk/pr/library/2011/01/18results.html (2011). Accessed 10 Apr 2011

BBC: Internet access is a fundamental human right. BBC News. Available at: http://news.bbc.co.uk/1/hi/technology/8548190.stm (2010). Accessed 10 Apr 2011

Doddington, G., Liggett, W., Martin, A., Przybocki, M., Reynolds, D.: Sheep, goats, lambs and wolves: a statistical analysis of speaker performance in NIST 1998 Speaker recognition evaluation. In: Proceedings of the 5th International Conference on Spoken Language Processing (ICSLP), Sydney, Australia (1998)

Hocking, C., Furnell, S.M., Clarke, N.L., Reynolds, P.: A distributed and cooperative user authentication framework. In: Proceedings of the 6th International Conference on Information Assurance and Security (IAS 2010), Atlanta (2010)

Index

N. Clarke, *Transparent User Authentication: Biometrics, RFID and Behavioural Profiling*, DOI 10.1007/978-0-85729-805-8, © Springer-Verlag London Limited 2011

About the Author

Dr. Nathan Clarke is an associate professor of information security and digital forensics at the Plymouth University in the United Kingdom and an adjunct associate professor with Edith Cowan University in Western Australia. He has been active in research since 2000, with interests in biometrics, mobile security, intrusion detection, digital forensics and information security awareness. An underlying theme to his research has been the focus upon transparent authentication, the underlying biometric techniques and associative system-level issues.

Dr. Clarke is a chartered engineer, a fellow of the British Computer Society, and a senior member of the IEEE. He is also active as a UK representative in International Federation for Information Processing (IFIP) working groups relating to the Human Aspects of Information Security and Assurance (which he co-created and currently acts as the vice-chair), Information Security Management, Information Security Education and Identity Management.

During his academic career, Dr. Clarke had authored over 60 publications in referred international journals and conferences. He is the current co-chair of the *Human Aspects of Information Security & Assurance (HAISA)* symposium and of the *Workshop on Digital Forensics and Incident Analysis (WDFIA)*. Dr. Clarke has also served on over 40 international conference events and regularly acts as a reviewer for numerous journals including, *Computers & Security*, *IEEE Transactions on Information Forensics and Security*, *The Computer Journal* and *Security & Communication Networks*. Dr. Clarke is the author of *Computer Forensics: A Pocket Guide* published by IT Governance.

N. Clarke, *Transparent User Authentication: Biometrics, RFID and Behavioural Profiling*,
DOI 10.1007/978-0-85729-805-8, © Springer-Verlag London Limited 2011

MIX
Papier aus verantwortungsvollen Quellen
Paper from responsible sources
FSC® C105338

If you have any concerns about our products,
you can contact us on
ProductSafety@springernature.com

In case Publisher is established outside the EU,
the EU authorized representative is:
Springer Nature Customer Service Center GmbH
Europaplatz 3, 69115 Heidelberg, Germany

Printed by Libri Plureos GmbH
in Hamburg, Germany